Lecture Notes in Computer Science　　　10151

Commenced Publication in 1973
Founding and Former Series Editors:
Gerhard Goos, Juris Hartmanis, and Jan van Leeuwen

More information about this series at http://www.springer.com/series/7409

Andrea Calì · Dorian Gorgan
Martín Ugarte (Eds.)

Semantic Keyword-Based Search on Structured Data Sources

COST Action IC1302
Second International KEYSTONE Conference, IKC 2016
Cluj-Napoca, Romania, September 8–9, 2016
Revised Selected Papers

Springer

Editors
Andrea Calì
Department of Computer Science and
  Information Systems
Birkbeck University of London
London
UK

Dorian Gorgan
Computer Science Department
Technical University of Cluj-Napoca
Cluj-Napoca
Romania

Martín Ugarte
Computer and Decision Engineering (CoDE)
  Department
Université Libre de Bruxelles
Brussels
Belgium

ISSN 0302-9743   ISSN 1611-3349   (electronic)
Lecture Notes in Computer Science
ISBN 978-3-319-53639-2   ISBN 978-3-319-53640-8   (eBook)
DOI 10.1007/978-3-319-53640-8

Library of Congress Control Number: 2017931541

LNCS Sublibrary: SL3 – Information Systems and Applications, incl. Internet/Web, and HCI

Printed on acid-free paper

This Springer imprint is published by Springer Nature
The registered company is Springer International Publishing AG
The registered company address is: Gewerbestrasse 11, 6330 Cham, Switzerland

Preface

In data management we face the problem of handling and querying very large datasets with large and partially unknown schemas, possibly containing billions of instances. An important issue in this context is to efficiently perform keyword searches. The size of the datasets poses challenges related to both scalability and semantic analysis. Another challenge is the discovery of suitable data sources for keyword search, given that one would want to process queries on relevant sources.

The Second International KEYSTONE Conference (IKC 2016), organized within the Cost Action IC1302 (Semantic Keyword-Based Search on Structured Data Sources), attracted several contributions in the area of keyword and semantic search on large structured data. In all, 14 papers were selected; the topics covered, among others, the areas of keyword extraction, natural language searches, graph databases, information retrieval techniques for keyword search and document retrieval. The program also included invited talks by experts in the field: Maria-Esther Vidal (University of Bonn, Germany), Dan Olteanu (University of Oxford, UK), Mihai Dinsoreanu (Recognos, Romania), Stefan Dietze (LS3 Research Centre, University of Hannover, Germany), Dragan Ivanovic (University of Novi Sad, Serbia), and Radu Tudoran (Huawei, Germany). An exciting panel moderated by Yannis Velegrakis took place with the participation of Radu Tudoran and Vagan Terziyan. The program was stimulating and managed to keep the participants in the lecture room despite the wonderful sights of Cluj-Napoca.

The success of this conference is the result of the effort of many. We would like to thank the authors, the invited speakers, the conference participants, the members of the Program Committee, and the external referees. We would also like to thank Springer for providing assistance in the preparation of the proceedings, the University of Cluj-Napoca for providing local facilities, and the local organizers and students who helped run the event.

COST (European Cooperation in Science and Technology) is a pan-European intergovernmental framework. Its mission is to enable break through scientific and technological developments leading to new concepts and products and thereby contribute toward strengthening Europe's research and innovation capacities. It allows researchers, engineers, and scholars to jointly develop their own ideas and take new initiatives across all fields of science and technology, while promoting multi- and interdisciplinary approaches. COST aims at fostering a better integration of less research-intensive countries to the knowledge hubs of the European research area. The COST Association, an international not-for-profit association under Belgian law, integrates all management, governing, and administrative functions necessary for the operation of the framework. The COST Association currently has 36 member countries.

September 2016

Andrea Calì
Dorian Gorgan
Martín Ugarte

Organization

IKC 2016 was organized within the Cost Action 1302 (Semantic Keyword-Based Search on Structured Data Sources), by the Computer Science Department, Faculty of Automation and Computer Science of the Technical University of Cluj-Napoca.

General Chair

Riccardo Torlone   Università Roma Tre, Italy

Program Chairs

Andrea Calì   Birkbeck University of London, UK
Dorian Gorgan   Technical University of Cluj-Napoca, Romania.
Martín Ugarte   Université Libre de Bruxelles, Belgium

Organizing Chairs

Dorian Gorgan   Technical University of Cluj-Napoca, Romania
Victor Bacu   Technical University of Cluj-Napoca, Romania

Invited Speakers

Maria-Esther Vidal   University of Bonn, Germany
Dan Olteanu   University of Oxford, UK
Dragan Ivanovic   University of Novi Sad, Serbia
Radu Tudoran   Huawei Research Engineer, Germany

Additional Reviewers

V. Alexiev
R. Amaro
I. Anagnostopoulos
M. Bielikova
C. Bobed
F. Bobillo
K. Belhajjame
E. Domnori
M. Dragoni

J. Espinoza
F. Guerra
S. Ilarri
E. Ioannou
D. Ivanovic
A. Kovacevic
M. López Nores
J. Lacasta
M. Lupu

F. Mandreoli   F. Pop
A. Mestrovic   V. Stoykova
P. Missier   G. Vargas-Solar
A. Nuernberger   L. Vintan

Sponsoring Institutions

COST: European Cooperation in Science and Technology (www.cost.eu)

Contents

Invited Papers

Retrieval, Crawling and Fusion of Entity-centric Data on the Web

Stefan Dietze[✉]

L3S Research Center, Leibniz Universität Hannover, Hannover, Germany
`dietze@L3S.de`

Abstract. While the Web of (entity-centric) data has seen tremendous growth over the past years, take-up and re-use is still limited. Data vary heavily with respect to their scale, quality, coverage or dynamics, what poses challenges for tasks such as entity retrieval or search. This chapter provides an overview of approaches to deal with the increasing heterogeneity of Web data. On the one hand, recommendation, linking, profiling and retrieval can provide efficient means to enable discovery and search of entity-centric data, specifically when dealing with traditional knowledge graphs and linked data. On the other hand, embedded markup such as Microdata and RDFa has emerged a novel, Web-scale source of entity-centric knowledge. While markup has seen increasing adoption over the last few years, driven by initiatives such as schema.org, it constitutes an increasingly important source of entity-centric data on the Web, being in the same order of magnitude as the Web itself with regards to dynamics and scale. To this end, markup data lends itself as a data source for aiding tasks such as knowledge base augmentation, where data fusion techniques are required to address the inherent characteristics of markup data, such as its redundancy, heterogeneity and lack of links. Future directions are concerned with the exploitation of the complementary nature of markup data and traditional knowledge graphs.

Keywords: Entity retrieval · Dataset recommendation · Markup · Schema.org · Knowledge graphs

1 Introduction

The emergence and wide-spread use of knowledge graphs, such as Freebase [5], YAGO [31], or DBpedia [1] as well as publicly available linked data [2], has led to an abundance of entity-centric available on the Web. Data is shared as part of datasets, usually containing interdataset links [23], which link equivalent, similar or related entities, while the majority of these links are concentrated on established reference graphs [12]. Datasets vary significantly with respect to represented resource types, currentness, coverage of topics and domains, size, used languages, coherence, accessibility [7] or general quality aspects [16]. Also, while entity-centric knowledge bases capture large amounts of factual knowledge in the form of RDF triples (subject-predicate-object), they still are

© Springer International Publishing AG 2017
A. Calì et al. (Eds.): IKC 2016, LNCS 10151, pp. 3–16, 2017.
DOI: 10.1007/978-3-319-53640-8_1

incomplete and inconsistent [39], i.e., coverage, quality and completeness vary heavily across types or domains, where in particular long-tail entities usually are insufficiently represented. In addition, while sharing of vocabularies and vocabulary terms is a crucial requirement for enabling reuse, the Web of data still features a large amount of highly overlapping and often unlinked vocabularies [8].

The wide variety and heterogeneity of available data(sets) and their characteristics pose significant challenges for data consumers when attempting to find and reuse useful data without prior knowledge about the available data and their features. This is seen as one of the reasons for the strong bias towards reusing well-understood reference graphs like Freebase or Yago, while there exists a long tail of datasets which is hardly considered or reused by data consumers.

In this work, we discuss a range of research results which aim at improving search and retrieval of entity-centric Web data. In Sect. 2, we will focus specifically on approaches towards dealing with the aforementioned heterogeneity of available knowledge graphs and linked data by introducing methods for recommendation and profiling of datasets as well as for enabling efficient entity retrieval. In Sect. 3, we will look beyond traditional linked data and discuss new forms of emerging entity-centric data on the Web, namely structured Web markup annotations embedded in HTML pages. Markup data has become prevalent on the Web, building on standards such as RDFa[1], Microdata[2] and Microformats[3], and driven by initiatives such as schema.org, a joint effort led by Google, Yahoo!, Bing and Yandex. We will introduce a number of case studies about scope and coverage of Web markup and introduce recent research which aims at exploiting Web markup data for tasks such as knowledge base population, data fusion or entity retrieval.

While this paper aims at providing a subjective overview of recent works as well as current and future research directions for the exploitation of entity-centric Web data, it is worthwhile to highlight that an exhaustive survey is beyond the scope of this paper.

2 Recommendation, Profiling and Retrieval of Entity-centric Web Data

The growth of structured linked data on the Web covers cross-domain and domain-specific data from a wide range of domains, where bibliographic (meta)data, such as [10], general resource metadata [32] and data from the life sciences domain [11] are particularly well represented. However, reuse of both vocabularies [8] as well as data is still limited. Particularly with respect to interlinking, the current topology of the linked data Web graph underlines the need for practical and efficient means to recommend suitable datasets: currently, only few, well established knowledge graphs show a high amount of inlinks, with

---

[1] RDFa W3C recommendation: http://www.w3.org/TR/xhtml-rdfa-primer/.

[2] http://www.w3.org/TR/microdata.

[3] http://microformats.org.

DBpedia being the most obvious target [30], while a long tail of datasets is largely ignored.

To facilitate search and reuse of existing datasets, descriptive and reliable metadata is required. However, as witnessed in the popular dataset registry DataHub[4], dataset descriptions often are missing entirely, or are outdated, for instance describing unresponsive endpoints [7]. This issue is partially due to the lack of automated mechanisms for generating reliable and up-to-date dataset metadata and hinders the reuse of datasets. The dynamics and frequent evolution of data further exacerbates this problem, calling for scalable and frequent update mechanisms of respective metadata.

In this section, we will introduce approaches aiming at facilitating data reuse and retrieval through (a) automated means for dataset recommendation, (b) dataset profiling as a means to facilitate dataset discovery through generating descriptive dataset metadata, and (c) improved entity retrieval techniques which address the heterogeneity of Web data, particularly, the prevalent lack of explicit links.

2.1 Dataset Recommendation

Dataset recommendation approaches such as [20,27] or the more recent works [12,13] tackle the problem of computing a ranking of datasets of relevance for the linking task, i.e. likely to contain linking candidates for a given source dataset. Formally speaking, dataset recommendation considers the problem of computing a rank score for each elements of a set of target datasets D_T so that the rank score indicates the relatedness of D_T to a given source dataset.

Leme *et al.* [19] present a ranking method based on Bayesian criteria and on the popularity of the datasets, what affects the applicability of the approach. The authors address this drawback in [20] by exploring the correlation between different sets of features, such as properties, classes and vocabularies.

Motivated by the observation that datasets often reuse vocabulary terms, [13] adopts the notion of a dataset profile, defined as a set of concept labels that describe the dataset and propose the *CCD-CosineRank* dataset recommendation approach, based on schema similarity across datasets. The approach consists of identifying clusters of comparable datasets, and, ranking the datasets in each cluster with respect to a given dataset. For the latter step, three different similarity measures are considered and evaluated. The approach is applied to the real-world datasets from the Linked Open Data graph and compared to two baseline methods, where results show a mean average precision of around 53% for recall of 100%, which indicates that *CCD-CosineRank* can reduce considerably the cost of dataset interlinking. As a by-product, the system returns sets of schema concept mappings between source and target datasets.

However, next to schema-level features, consideration of instance-level characteristics is crucial when computing overlap and complementarity of described entities. Given the scale of available datasets, exhaustive comparisons of schemas

---

[4] http://www.datahub.io.

and instances or some of their features are not feasible as an online process. For instance, in [14] (Sect. 2.3), authors generate a weighted bipartite graph, where datasets and topics represent the nodes, related through weighted edges, indicating the relevance of a topic for a specific dataset. While computation of such topic profiles is costly, it is usually applied to a subset of existing datasets only, where any new or so far unannotated datasets require the pre-computation of a dedicated topic profile.

[12] builds on this observation and provides a recommendation method which not only takes into account the direct relatedness of datasets as emerging from the topic-dataset graph produced through the profiling in [14], but also adopts established collaborative filtering (CF) practices by considering the topic relationships emerging from the global topic-dataset-graph to derive specific dataset recommendations. CF enables to consider arbitrary (non-profiled) datasets as part of recommendations. This approach on the one hand significantly increases the recall of recommendations, and at the same time improves recommendations through considering dataset connectivity as another relatedness indicator. The intuition is that global topic connectivity provides reliable connectivity indicators even in cases where the underlying topic profiles might be noisy, i.e. that, even poor or incorrect topic annotations will serve as reliable relatedness indicator when shared among datasets. Theoretically, this approach is agnostic to the underlying topic index. This approach also reflects both, instance-level as well as schema-level characteristics of a specific dataset. Even though topics are derived from instances, resources of particular types show characteristic topic distributions, which significantly differ across different types [34].

Applied to the set of all available linked dataasets, experiments show superior performance compared to three simple baselines, namely based on shared key-words, shared topics, and shared common links, achieving a reduction of the original search space of up to 86% on average. It is worth to highlight that the aforementioned evaluation results are affected by the limited nature of available ground truth data, where all works relied on linkset descriptions from the DataHub. However, while this data is manually curated, it is inherently sparse and incomplete, that is, providers usually indidate a very limited amount of linking information. This leads to inflated recall values and at the same time, affects precision in the sense that results tend to label correct matches as false positives according to the ground truth. One future direction of research aims at producing a more complete ground truth. Given the scale of available data on the Web, computing linking metrics should resort to sampling and approximation strategies.

2.2 Dataset Profiling

Rather than automatically recommending datasets, additional metadata can enable data consumers to make an informed decision when selecting suitable datasets for a given task. In [14], authors address this challenge of automatically extracting dataset metadata with the goal of facilitating dataset search and reuse. Authors propose an approach for creating structured dataset profiles, where a profile describes the topic coverage of a particular dataset through a weighted graph of selected DBpedia categories. The approach consists of a processing pipeline that combines tailored techniques for dataset sampling, topic extraction from reference datasets and topic relevance ranking. Topics are extracted through named entity recognition (NER) techniques and the use of a reference category vocabulary, namely DBpedia. Relevance of a particular category for a dataset is computed based on graphical models like *PageRank* [6], *K-Step Markov* [36], and *HITS* [18]. While this is a computationally expensive process, authors experimentally identify the parameters which enable a suitable trade-off between representativeness of generated profiles and scalability. Finally, generated dataset profiles are exposed as part of a public structured dataset catalog based on the *Vocabulary of Interlinked Datasets* (VoID[5]) and the recent vocabulary of links (VoL)[6].

As part of the experiments, authors generated dataset profiles for all accessible linked datasets classified as Linked Open Data on the DataHub and demonstrate that, even with comparably small sample sizes (10%), representative profiles and rankings can be generated. For instance, ΔNDCG=0.31 is achieved when applying *KStepM* and an additional normalisation step. The results demonstrate superior performance when compared to *LDA* with ΔNDCG=0.10 applied to complete set of resource instances. The main contribution consists of (i) a scalable method for efficiently generating structured dataset topic profiles combining and configuring suitable methods for NER, topic extraction and ranking as part of an experimentally optimised configuration, and (ii) the generation of structured dataset profiles for a majority of linked datasets according to established dataset description vocabularies. Dataset profiles generated through this approach can be explored in a stand-alone online explorer[7], top-k topic annotations are used as part of the LinkedUp dataset catalog [8], and more recently, topic profiles have been used to develop dataset recommendation techniques [12]. While it has been noted that meaningfulness and comparability of topic profiles can be increased when considering topics associated with certain resource types only, as part of additional work, resource type-specific dataset profiling approaches have been introduced [34].

2.3   Improving Entity Retrieval

While previous sections address the problem of discovering datasets, i.e. graphs representing potentially large amounts of entities, the entity-centric nature of the Web of data involves tasks related to entity and object retrieval [3,35] or entity-driven text summarization [9]. Major search engine providers such as Google and Yahoo! already exploit entity-centric data to facilitate semantic search using knowledge graphs. In such scenarios, data is aggregated from a range of sources calling for efficient means to search and retrieve entities in large data graphs.

---

[5] http://vocab.deri.ie/void.

[6] http://data.linkededucation.org/vol/.

[7] http://data-observatory.org/lod-profiles/profile-explorer/.

In particular, *entity retrieval* (also known as *Ad-Hoc Object retrieval*) [26,35] aims at retrieving relevant entities given a particular entity-seeking query, resulting in a ranked list of entities [3]. By applying standard information retrieval algorithms, like BM2F, on constructed indexes over the textual descriptions (*literals*) of entities, previous works have demonstrated promising performance.

While there is a large amount of queries that are topic-based, e.g. 'U.S. Presidents', rather than entity-centric, approaches like [35] have proposed retrieval techniques that make use of explicit links between entities for result or query expansion, for instance, *owl:sameAs* or *rdfs:seeAlso* statements. However, such statements are very sparse, particularly across distinct datasets.

[15] proposes a method for improving entity retrieval results by *expanding* and *re-ranking* the result set from a baseline retrieval model (BM25F). Link sparsity is addressed through clustering of entities (*x-means* and *spectral* clustering), based on their similarity, using both lexical and structural features. Thus, result sets retrieved through the baseline approach are expanded with related entities residing the same clusters as the result set entities. Subsequent re-ranking considers the similarity to the original query, and their relevance likelihood based on the corresponding entity type, building on the assumption that particulary query types are more likely result in certain result types (*query type affinity*). The clustering process is carried out as offline preprocessing, while the entity retrieval, expansion and re-ranking are performed online. An experimental evaluation on the BTC12 dataset [17], where the clustering process was carried out on a large set of entities (over 450 million), and using the SemSearch[8] query dataset shows that the proposed approach outperforms existing basslines significantly.

3   Crawling and Fusion of Entity-centric Web Markup

While the previous section has discussed approaches for exploiting entity-centric data from traditional knowledge graphs and linked data, here we turn towards structured Web markup as an emerging and unprecedented source of entity-centric Web data. Markup annotations embedded in HTML pages have become prevalent on the Web, building on standards such as RDFa[9], Microdata[10] and Microformats[11], and driven by initiatives such as schema.org, a joint effort led by Google, Yahoo!, Bing and Yandex.

The Web Data Commons [22], a recent initiative investigating a Web crawl of 2.01 billion HTML pages from over 15 million pay-level-domains (PLDs) found that 30% of all pages contain some form of embedded markup already, resulting in a corpus of 20.48 billion RDF quads[12]. The scale and upward trend of adoption[13] - the proportion of pages containing markup increased from 5.76%

---

[8] http://km.aifb.kit.edu/ws/semsearch10/.

[9] RDFa W3C recommendation: http://www.w3.org/TR/xhtml-rdfa-primer/.

[10] http://www.w3.org/TR/microdata.

[11] http://microformats.org.

[12] http://www.webdatacommons.org.

[13] http://webdatacommons.org/structureddata/.

to 30% between 2010 and 2014 - suggest potential for a range of tasks, such as entity retrieval and knowledge base augmentation. However, facts extracted from embedded markup have different characteristics when compared to traditional knowledge graphs and linked data. In the following, we discuss first some case studies which investigate the coverage and distribution of Web markup for a particular set of entity types (Sect. 3.1), then we discuss apparent challenges (Sect. 3.2), and finally, we introduce current research which apply data fusion techniques to use Web markup data in the aforementioned tasks (Sect. 3.3).

3.1 Case Studies: Type-Specific Coverage of Web Markup

As part of type-specific investigations [28, 29], we have investigated the scope, distribution and coverage of Web markup, specifically considering the cases of bibliographic data and of learning resource annotations. [28] provides a study of the adoption of Web markup for the annotation of bibliographic entities, being the first effort to investigate scholarly data extracted from embedded annotations. Utilising the WDC as largest crawl of embedded markup so far, the investigation considers all statements which describe entities (subjects) that are of type *s:ScholarlyArticle* or of any type but co-occurring on the same document with any *s:ScholarlyArticle* instance. Here and in the following we refer to the http://schema.org namespace as *s:*, and abbreviate *s:ScholarlyArticle* as *s:SchoArt*. Although there is a wide variety of types used for bibliographic and scholarly information, *s:SchoArt* is the only type which explicitly refers to scholarly bibliographic data. While this is a limitation with respect to recall, we followed this approach to enable a high precision of the analysed data within the scope of our study.

The extracted dataset contains 6,793,764 quads, 1,184,623 entities, 83 distinct classes, and 429 distinct predicates. Insights are provided with respect to frequent data providers, the adoption and usage of terms and the distribution across providers, domains and topics. The distribution of extracted data, spread across 214 distinct Pay-Level-Domains (PLDs), 38 Top-Level-Domains (TLDs) and 199,980 documents is represented in Fig. 1. The blue (lower) line corresponds to the distribution of entities and the red and dashed (upper) line corresponds to the distribution of statements over PLDs/TLDs and documents. The number of entities/statements presented on the *y-axis* are plotted in the logarithmic scale. An apparent observation is the power law-like distribution, where usually a small amount of sources (PLDs, TLDs, documents) provide the majority of entities and statements. For example *springer.com* alone exposes a total of 850,697 entities and 3,011,702 statements. The same pattern can be identified for vocabulary terms, where few predicates are highly used, complemented by a long tail of predicates of limited use. With regard to the distribution across top-level-domains, a certain bias towards French data providers seems apparent based on some manual investigation of the top-k genres and publishers. Article titles, PLDs and publishers suggest a bias towards specific disciplines, namely Computer Science and the Life Sciences which mirrors a similar pattern in the linked data world. However, the question to what extent this is due to the

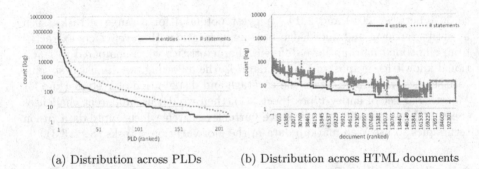

(a) Distribution across PLDs   (b) Distribution across HTML documents

Fig. 1. Distribution of entities/statements over PLDs and documents (from [28]). (Color figure online)

selective content of the Common Crawl or representative for schema.org adoption on the Web in general requires additional investigations.

[33] investigates the same corpus, yet towards the goal of understanding the adoption of *LRMI*[14] statements. The *Learning Resources Metadata Initiative (LRMI)* provides a schema.org extension tailored to the annotation of educational resources. In order to assess not only the coverage but also the evolution of LRMI statements, authors extracted subsets from the WDC2013 and WDC2014 datasets, by selecting all quads which co-occur with any of the LRMI vocabulary terms, such as *educationalAlignment, educationalUse, timeRequired*, or *typicalAgeRange*. The subsets under investigation contain 51,601,696 (WDC2013) respectively 50,901,532 (WDC2014) quads. The total number of entities in 2013 is 10,469,565 while in 2014 there are 11,861,807 entities, showing a significant growth in both cases. Regarding documents, we observe 3,060,024 documents in 2013 and 4,343,951 in 2014. Similarly to the case of bibliographic data, the distribution follows a power-law, where a small amount of providers (PLDs) provide large proportions of the data.

Findings from both studies suggest an uneven distribution of quads across documents and providers leading to potential bias in obtained entity-centric knowledge. On the other hand, the studies provide first evidence of a widespread adoption of even domain-specific types and terms, where in both cases, an inspection of the PLDs suggest that key data providers, such as publishers, libraries, or journals already embrace Web markup for improving search and interpretation of their Web pages. More exhaustive studies should consider, however, the use of focused crawls, which enable a more comprehensive study into the adoption of markup annotations in a respective domain.

3.2   Challenges

Initial investigations have shown the complementary nature of markup data, when compared to traditional knowledge bases, both at the entity level as well as

---

[14] http://www.lrmi.net.

the fact level, where the extent of additional information varies strongly between resource types. Though Web markup constitutes a rich and dynamic knowledge resource, the problem of answering entity-centric queries from entity descriptions extracted from embedded markup is a novel challenge, where the specific characteristics of such data pose different challenges [37] compared to traditional linked data:

- **Coreferences**: entities, particularly popular ones, are represented on a multitude of pages, resulting in vast amounts co-referring entity descriptions about the same entity. For instance, 797 entity descriptions can be obtained from WDC2014 which are of type *s:movie* and show a label (*s:name*) *Forrest Gump*.
- **Lack of explicit links**: RDF statements extracted from markup form a very sparsely linked graph, as opposed to the higher connectivity of traditional RDF datasets. This problem is elevated by the large amount of coreferences, where explicit links would facilitate the fusion of facts about the same entity from a variety of sources.
- **Redundant statements**: extracted RDF statements are highly redundant. For instance, Fig. 2 presents a power law distribution for predicates observed from entity descriptions of type *s:Movie* and *s:Book*, where a few popular predicates occur in the vast majority of statements, followed by a long tail of infrequent predicates. Authors also observe that only a small proportion of facts are lexically distinct (60%), many of which are near-duplicates.
- **Errors**: as documented in [21], data extracted from markup contains a wide variety of syntactic and semantic errors, including typos or the misuse of vocabulary terms.

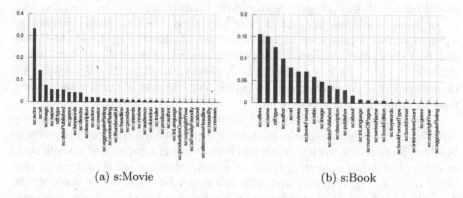

(a) s:Movie   (b) s:Book

Fig. 2. Statement distribution across predicates for types *s:Movie* and *s:Book* (from [37]).

To this end, entity descriptions complement each other, yet sophisticated data fusion techniques are required in order to enable further exploitation of entity-centric knowledge from Web markup.

3.3 Exploiting Web Markup: Data Fusion and Knowledge Base Augmentation

Initial works such as the *Glimmer* search engine[15] have applied traditional *entity retrieval* techniques [4] to embedded markup (WDC). However, given the large amount of flat and highly redundant entity descriptions, practical use of search results obtained in that way is limited [37]. Key issues of such approaches include identity resolution as well as the vast amount of duplicates and near-duplicates. Therefore, the application of *data fusion* techniques is required to obtain a consolidated and correct entity description when answering entity-centric queries.

However, given the dynamics of Web markup data, the validity and correctness of a fact is usually of temporal nature [24]. For instance, the predicate *s:price* of a particular product offer is highly dynamic and its correctness depends strongly on the considered time frame. For these reasons, any data fusion approach would have to consider efficiency in order to enable frequent repetitions of the extraction pipeline, consisting of crawling, extraction and fusion. Considering the scale of large Web crawls such as the WDC, general data fusion strategies which are applied over the entire pool of data are impractical. This suggests a need for focused approaches, which are able to efficiently obtain fused entity descriptions for a given set of entity-seeking queries. For instance, for an entity-seeking query *'Iphone 6'* of type product, a *query-centric data fusion* approach will fuse all correct facts from an available corpus or crawl into a diverse entity description.

In [38], authors present *Clustering-Based Fact Selection(CBFS)* as an approach for query-centric data fusion of Web markup. Entity retrieval is conducted to provide a set of candidate facts for a given query. For this purpose, authors build a standard IR index of entity descriptions and apply the BM25 retrieval model on pseudo-key properties to obtain candidate entity descriptions. One major issue to address in the fact selection process, is the canonicalization of different surface forms, such as `Tom Hanks` and `T. Hanks`. To detect duplicates and near duplicates, authors cluster entity labels at the predicate level into n clusters $(c_1, c_2, \cdots, c_n) \in C$ using the X-Means algorithm [25]. Fact selection then considers a set of heuristics to enable the selection of correct and diverse facts from the candidate pool.

Experiments using the WDC2014 dataset indicate a comparably high precision 83.3% of this initial approach, showing a gain of 5.5% compared to a simple baseline. More recent work is concerned with building a supervised classification model for the data fusion step, based on a comprehensive feature set which considers relevance, quality and authority of sources, facts and entity descriptions. To evaluate the potential of this approach for aiding knowledge base augmentation tasks, authors also measure the coverage gain by comparing obtained entity descriptions to their corresponding descriptions in DBpedia. It was found that 57% of the facts detected by CBFS do not exist in DBpedia with some of the facts corresponding to new predicates and some to already existing ones, which are

---

[15] http://glimmer.research.yahoo.com/.

not sufficiently populated. Considering only the predicates that exist in DBpedia, and the coverage gain is 33.4%. Currently ongoing research addresses the use of Web markup for tasks such knowledge base augmentation and temporal entity interlinking.

4   Conclusions

This paper provided an overview on selected works on retrieval, crawling and fusion of entity-centric Web data. While the heterogeneity and diversity of traditional linked data and knowledge graphs calls for efficient methods for dataset recommendation, profiling or entity retrieval (Sect. 2), we also investigated the exploitation of embedded Web markup data as emerging form of large-scale entity-centric data on the Web (Sect. 3). While an exhaustive literature review is out of scope of this paper, the focus here is on selected works covering a range of topics of relevance to general aim of retrieving entity-centric data from the Web.

Promising future directions are specifically concerned with the convergence of both sources of entity-centric knowledge discussed in this paper, for instance, by exploiting Web markup and data from Web tables for knowledge base augmentation. Interesting opportunities also emerge from the large-scale availability of markup and its use as unprecedented source of training data for supervised entitiy recognition, disambigutation or interlinking methods. The availability of explicit entity annotations at Web-scale enables the computation of a wide range of features which consider both, characteristics of unstructured Web documents as well as the embedded entity markup.

Acknowledgements. While all discussed works are joint research with numerous colleagues, friends and collaborators from a number of research institutions, the author would like to thank all involved researchers for the inspiring and productive work throughout the previous years. In addition, the author expresses his gratitude to all funding bodies that enabled the presented research through a variety of funding programs.

References

1. Auer, S., Bizer, C., Kobilarov, G., Lehmann, J., Cyganiak, R., Ives, Z.: DBpedia: a nucleus for a web of open data. In: Aberer, K., Choi, K.-S., Noy, N., Allemang, D., Lee, K.-I., Nixon, L., Golbeck, J., Mika, P., Maynard, D., Mizoguchi, R., Schreiber, G., Cudré-Mauroux, P. (eds.) ASWC/ISWC -2007. LNCS, vol. 4825, pp. 722–735. Springer, Heidelberg (2007). doi:10.1007/978-3-540-76298-0_52
2. Bizer, C., Heath, T., Berners-Lee, T.: Linked data - the story so far. Int. J. Semantic Web Inf. Syst. **5**(3), 1–22 (2009)
3. Blanco, R., Cambazoglu, B.B., Mika, P., Torzec, N.: Entity recommendations in web search. In: Alani, H., Kagal, L., Fokoue, A., Groth, P., Biemann, C., Parreira, J.X., Aroyo, L., Noy, N., Welty, C., Janowicz, K. (eds.) ISWC 2013. LNCS, vol. 8219, pp. 33–48. Springer, Heidelberg (2013). doi:10.1007/978-3-642-41338-4_3

4. Blanco, R., Mika, P., Vigna, S.: Effective and efficient entity search in RDF data. In: Aroyo, L., Welty, C., Alani, H., Taylor, J., Bernstein, A., Kagal, L., Noy, N., Blomqvist, E. (eds.) ISWC 2011. LNCS, vol. 7031, pp. 83–97. Springer, Heidelberg (2011). doi:10.1007/978-3-642-25073-6_6

5. Bollacker, K., Evans, C., Paritosh, P., Sturge, T., Taylor, J.: Freebase: a collaboratively created graph database for structuring human knowledge. In: Proceedings of the 2008 ACM SIGMOD International Conference on Management of Data. SIGMOD 2008, pp. 1247–1250. ACM, New York (2008)

6. Brin, S., Page, L.: The anatomy of a large-scale hypertextual web search engine. Comput. Netw. **30**(1–7), 107–117 (1998)

7. Buil-Aranda, C., Hogan, A., Umbrich, J., Vandenbussche, P.-Y.: SPARQL web-querying infrastructure: ready for action? In: Alani, H., Kagal, L., Fokoue, A., Groth, P., Biemann, C., Parreira, J.X., Aroyo, L., Noy, N., Welty, C., Janowicz, K. (eds.) ISWC 2013. LNCS, vol. 8219, pp. 277–293. Springer, Heidelberg (2013). doi:10.1007/978-3-642-41338-4_18

8. DAquin, M., Adamou, A., Dietze, S.: Assessing the educational linked data landscape. In: ACM Web Science 2013 (WebSci 2013), Paris, France. ACM (2013)

9. Demartini, G., Missen, M.M.S., Blanco, R., Zaragoza, H.: Entity summarization of news articles. In: Proceedings of the 33rd ACM SIGIR, pp. 795–796 (2010)

10. Dietze, S., Taibi, D., dAquin, M.: Facilitating scientometrics in learning analytics and educational data mining - the LAK dataset. Semantic Web J. **8**(3), 395–403 (2017)

11. Dietze, S., Taibi, D., Yu, H.Q., Dovrolis, N.: A linked dataset of medical educational resources. Br. J. Educ. Technol. BJET **46**(5), 1123–1129 (2015)

12. Ben Ellefi, M., Bellahsene, Z., Dietze, S., Todorov, K.: Beyond established knowledge graphs-recommending web datasets for data linking. In: Bozzon, A., Cudre-Maroux, P., Pautasso, C. (eds.) ICWE 2016. LNCS, vol. 9671, pp. 262–279. Springer, Heidelberg (2016). doi:10.1007/978-3-319-38791-8_15

13. Ben Ellefi, M., Bellahsene, Z., Dietze, S., Todorov, K.: Dataset recommendation for data linking: an intensional approach. In: Sack, H., Blomqvist, E., d'Aquin, M., Ghidini, C., Ponzetto, S.P., Lange, C. (eds.) ESWC 2016. LNCS, vol. 9678, pp. 36–51. Springer, Heidelberg (2016). doi:10.1007/978-3-319-34129-3_3

14. Fetahu, B., Dietze, S., Pereira Nunes, B., Antonio Casanova, M., Taibi, D., Nejdl, W.: A scalable approach for efficiently generating structured dataset topic profiles. In: Presutti, V., d'Amato, C., Gandon, F., d'Aquin, M., Staab, S., Tordai, A. (eds.) ESWC 2014. LNCS, vol. 8465, pp. 519–534. Springer, Heidelberg (2014). doi:10.1007/978-3-319-07443-6_35

15. Fetahu, B., Gadiraju, U., Dietze, S.: Improving entity retrieval on structured data. In: Arenas, M., et al. (eds.) ISWC 2015. LNCS, vol. 9366, pp. 474–491. Springer, Heidelberg (2015). doi:10.1007/978-3-319-25007-6_28

16. Guéret, C., Groth, P., Stadler, C., Lehmann, J.: Assessing linked data mappings using network measures. In: Simperl, E., Cimiano, P., Polleres, A., Corcho, O., Presutti, V. (eds.) ESWC 2012. LNCS, vol. 7295, pp. 87–102. Springer, Heidelberg (2012). doi:10.1007/978-3-642-30284-8_13

17. Harth, A.: Billion Triples Challenge data set. http://km.aifb.kit.edu/projects/btc-2012/ (2012)

18. Kleinberg, J.M.: Authoritative sources in a hyperlinked environment. J. ACM **46**(5), 604–632 (1999)

19. Leme, L.A.P.P., Lopes, G.R., Nunes, B.P., Casanova, M.A., Dietze, S.: Identifying candidate datasets for data interlinking. In: Daniel, F., Dolog, P., Li, Q. (eds.) ICWE 2013. LNCS, vol. 7977, pp. 354–366. Springer, Heidelberg (2013). doi:10.1007/978-3-642-39200-9_29

20. Rabello Lopes, G., Paes Leme, L.A.P., Pereira Nunes, B., Casanova, M.A., Dietze, S.: Two approaches to the dataset interlinking recommendation problem. In: Benatallah, B., Bestavros, A., Manolopoulos, Y., Vakali, A., Zhang, Y. (eds.) WISE 2014. LNCS, vol. 8786, pp. 324–339. Springer, Heidelberg (2014). doi:10.1007/978-3-319-11749-2_25

21. Meusel, R., Paulheim, H.: Heuristics for fixing common errors in deployed *schema.org* microdata. In: Gandon, F., Sabou, M., Sack, H., d'Amato, C., Cudré-Mauroux, P., Zimmermann, A. (eds.) ESWC 2015. LNCS, vol. 9088, pp. 152–168. Springer, Heidelberg (2015). doi:10.1007/978-3-319-18818-8_10

22. Meusel, R., Petrovski, P., Bizer, C.: The webdatacommons microdata, RDFa and microformat dataset series. In: Mika, P., Tudorache, T., Bernstein, A., Welty, C., Knoblock, C., Vrandečić, D., Groth, P., Noy, N., Janowicz, K., Goble, C. (eds.) ISWC 2014. LNCS, vol. 8796, pp. 277–292. Springer, Heidelberg (2014). doi:10.1007/978-3-319-11964-9_18

23. Pereira Nunes, B., Dietze, S., Casanova, M.A., Kawase, R., Fetahu, B., Nejdl, W.: Combining a co-occurrence-based and a semantic measure for entity linking. In: Cimiano, P., Corcho, O., Presutti, V., Hollink, L., Rudolph, S. (eds.) ESWC 2013. LNCS, vol. 7882, pp. 548–562. Springer, Heidelberg (2013). doi:10.1007/978-3-642-38288-8_37

24. Oulabi, Y., Meusel, R., Bizer, C.: Fusing time-dependent web table data. In: Proceedings of the 19th International Workshop on Web and Databases, p. 3. ACM (2016)

25. Pelleg, D., Moore, A.W. et al.: X-means: extending k-means with efficient estimation of the number of clusters. In: ICML, pp. 727–734 (2000)

26. Pound, J., Mika, P., Zaragoza, H.: Ad-hoc object retrieval in the web of data. In: Proceedings of the 19th WWW, pp. 771–780 (2010)

27. Lopes, G.R., Leme, L.A.P.P., Nunes, B.P., Casanova, M.A., Dietze, S.: Recommending tripleset interlinking through a social network approach. In: Lin, X., Manolopoulos, Y., Srivastava, D., Huang, G. (eds.) WISE 2013. LNCS, vol. 8180, pp. 149–161. Springer, Heidelberg (2013). doi:10.1007/978-3-642-41230-1_13

28. Sahoo, P., Gadiraju, U., Yu, R., Saha, S., Dietze, S.: Analysing structured scholarly data embedded in web pages. April 2016

29. Sahoo, P., Gadiraju, U., Yu, R., Saha, S., Dietze, S.: Analysing structured scholarly data embedded in web pages. In: Proceedings of the 25th International Conference on World Wide Web Companion. International World Wide Web Conferences Steering Committee (2016)

30. Schmachtenberg, M., Bizer, C., Paulheim, H.: Adoption of the linked data best practices in different topical domains. In: Mika, P., Tudorache, T., Bernstein, A., Welty, C., Knoblock, C., Vrandečić, D., Groth, P., Noy, N., Janowicz, K., Goble, C. (eds.) ISWC 2014. LNCS, vol. 8796, pp. 245–260. Springer, Heidelberg (2014). doi:10.1007/978-3-319-11964-9_16

31. Suchanek, F.M., Kasneci, G., Weikum, G.: Yago: a core of semantic knowledge. In: Williamson, C.L., Zurko, M.E., Patel-Schneider, P.F., Shenoy, P.J. (eds) WWW, pp. 697–706. ACM, New York (2007)

32. Taibi, D., Chawla, S., Dietze, S., Marenzi, I., Fetahu, B.: Exploring ted talks as linked data for education. Brit. J. Educational Tech. **46**(5), 1092–1096 (2015)

33. Taibi, D., Dietze, S.: Towards embedded markup of learning resources on the web: An initial quantitative analysis of LRMI terms usage. In: Bourdeau, J., Hendler, J., Nkambou, R., Horrocks, I., Zhao, B.Y. (eds.) WWW (Companion Volume), pp. 513–517. ACM, New York (2016)

34. Taibi, D., Dietze, S., Fetahu, B., Fulantelli, G.: Exploring type-specific topic profiles of datasets: a demo for educational linked data. In: Horridge, M., Rospocher, M., van Ossenbruggen, J. (eds.) International Semantic Web Conference - Posters and Demos, vol. 1272. CEUR Workshop Proceedings, pp. 353–356. CEUR-WS.org (2014)

35. Tonon, A., Demartini, G., Cudré-Mauroux, P.: Combining inverted indices and structured search for Ad-hoc object retrieval. In: Proceedings of the 35th ACM SIGIR, pp. 125–134 (2012)

36. White, S., Smyth, P.: Algorithms for estimating relative importance in networks. In: Proceedings of the 9th ACM SIGKDD International Conference on Knowledge Discovery and Data Mining (SIGKDD), pp. 266–275 (2003)

37. Yu, R., Fetahu, B., Gadiraju, U., Dietze, S.: A survey on challenges in web markup data for entity retrieval. In: 15th International Semantic Web Conference (ISWC 2016) (2016)

38. Yu, R., Gadiraju, U., Zhu, X., Fetahu, B., Dietze, S.: Towards entity summarisation on structured web markup. In: Sack, H., Rizzo, G., Steinmetz, N., Mladenić, D., Auer, S., Lange, C. (eds.) ESWC 2016. LNCS, vol. 9989, pp. 69–73. Springer, Heidelberg (2016). doi:10.1007/978-3-319-47602-5_15

39. Yuan, W., Demidova, E., Dietze, S., Zhou, X.: Analyzing relative incompleteness of movie descriptions in the web of data: a case study. In: Horridge, M., Rospocher, M., van Ossenbruggen, J. (eds.) International Semantic Web Conference - Posters and Demos, vol. 1272. CEUR Workshop Proceedings, pp. 197–200. CEUR-WS.org (2014)

Data Multiverse: The Uncertainty Challenge of Future Big Data Analytics

Radu Tudoran[✉], Bogdan Nicolae, and Götz Brasche

Huawei German Research Center, Riesstraße 25, 80992 München, Germany
{radu.tudoran,bogdan.nicolae,goetz.brasche}@huawei.com

Abstract. With the explosion of data sizes, extracting valuable insight out of big data becomes increasingly difficult. New challenges begin to emerge that complement traditional, long-standing challenges related to building scalable infrastructure and runtime systems that can deliver the desired level of performance and resource efficiency. This vision paper focuses on one such challenge, which we refer to as the analytics uncertainty: with so much data available from so many sources, it is difficult to anticipate what the data can be useful for, if at all. As a consequence, it is difficult to anticipate what data processing algorithms and methods are the most appropriate to extract value and insight. In this context, we contribute with a study on current big data analytics state-of-art, the use cases where the analytics uncertainty is emerging as a problem and future research directions to address them.

Keywords: Big data analytics · Large scale data processing · Data access model · Data uncertainty · Approximate computing

1 Introduction

Data is the new natural resource. Its ingestion and processing leads to valuable insight that is transformative in all aspects of our world [12]. Science employs data-driven approaches to create complex models that explain nature in great detail, otherwise impossible to obtain using traditional analytical approaches by themselves. In industry, data science is an essential value generator: it leverages a variety of modern data sources (mobile sensors, social media, etc.) to understand customer behaviors and financial trends, which ultimately facilitates better business decisions.

Helped by the rise of cloud computing, an entire data market has emerged that enables users to share and consume massive amounts of data from a variety of distributed data sources. At the core of this data market is the so called *big data analytics* ecosystem: an ensemble of techniques and middleware specifically designed to interpret unstructured data on-the-fly (e.g. Spark [16], Flink [1], Storm [13], etc.). As more value is extracted out of the data market using big data analytics, the interest for collecting more data emerges, which ultimately unleashes a chain reaction. An evidence for this is already visible in overall

© Springer International Publishing AG 2017
A. Calì et al. (Eds.): IKC 2016, LNCS 10151, pp. 17–22, 2017.
DOI: 10.1007/978-3-319-53640-8_2

annual world network traffic: the end of 2016 will mark the beginning of the Zettabyte era (i.e. 1 ZB), with growth predictions showing 2.3 ZB by 2020 [2].

However, while there is an obvious advantage of access to more data in that more insight can be potentially extracted, this also introduces an unprecedented challenge: it is becoming increasingly difficult for the user to know in advance how useful the data is (either whole or parts of it) or what algorithm would work best. We call this the *analytics uncertainty*. Because of it, users are often trapped in a sequential trial-and-error process: apply data transformation, interpret result, adjust algorithm, repeat. Given the data market explosion, this is simply too slow to produce the desired results in a timely fashion. Therefore, a new way to reason about big data at scale is needed.

In this paper we study the current state of art from the analytics uncertainty perspective. Our contribution can be summarized as follows: (1) we discuss the state-of-art big data analytics ecosystem and its expressivity in terms of data processing mechanisms offered to the users; (2) we discuss a series of use case patterns that exhibit an inherent analytics uncertainty; (3) based on these patterns, we identify and discuss the limitations of the current approaches.

2   Background and Related Work

We summarize in this Section major classes of big data analytics approaches.

2.1   MapReduce-Like

Data-oriented batch programming models that separate the computation from its parallelization gained rapid popularity beginning with the MapReduce [7] paradigm. While convenient for the user due to automated parallelization by design, at runtime-level this introduces a significant level of complexity. One major contribution in this area is the *data-locality centered design*: the storage layer is co-located with the compute elements and exposes the data locations such that the computation can be scheduled close to the data. Using this approach, scalability is possible even on commodity hardware. Combined with the ease of use, this lead to a wide adoption of MapReduce in production environments.

2.2   In-Memory Generic Processing

Over time, MapReduce presented several limitations, both in terms of expressivity (formulating a solution by means of *map* and *reduce* is often difficult) and performance (high overhead for I/O despite co-location of compute and storage). To this end, a new generation of *in-memory big data analytics* is increasingly gaining popularity over MapReduce such as Spark, Flink. By making heavy use of in-memory data caching such engines minimizes the interactions with the storage layer, which further reduces I/O bottlenecks due to slow local disks, extra copies and serialization issues. To improve expressivity these engines facilitates the development of multi-step data pipelines using a directed acyclic graph

(DAG) as a runtime construct based on the high-level control flow of the application defined via the exposed API.

One illustrative example is Spark [16], which relies on two main parallel programming abstractions: (1) *resilient distributed datasets* (RDDs), a partitioned data structure hosting the data itself; and (2) parallel operations on the RDDs. RDD holds provenance information (referred to as *lineage*) and can be rebuilt in case of failures by partial recomputation from ancestor RDDs. RDD operations can be either *narrow* (i.e. each output partition can be computed from its corresponding input partition directly) or *wide* (i.e., each output partition is computed by combining pieces from many other input partitions). Wide transformations are particularly challenging because they involve complex all-to-all data exchanges, which may introduce overhead with respect to performance, scalability and resource utilization [10]. Furthermore, optimized broadcast of data is another important issue [11].

2.3 Stream Processing

Live data sources (e.g., web services, social and news feeds, sensors, etc.) are increasingly playing a critical role in big data analytics. By introducing an online dimension to data processing, they improve the reactivity and "freshness" of the results, which ultimately can potentially lead to better insights. As a consequence, *big data stream processing* saw a rapid rise recently. Storm [13] was one of the first low-level engines that introduced basic semantics in terms of bolts and sprouts. As the need for highly reactive processing grew, other more sophisticated approaches appeared such as S4 [9], MillWheel [3], Apache Flink, Apache APEX or Apache Samza. These approaches address issues such as fault tolerance, low latency and consistency guarantees (at least one, at most one, exactly one). A complementary effort towards low latency also gained significant traction: the idea of optimizing batch processing to handle frequent mini-batches. This approach was popularized by Spark via DStreams [17], then followed by Apache Storm Trident and data management tools like JetStream [14].

A significant effort to formalize stream computing in a way that captures the liveness of input data and the results by design is the *window* concept. Windows are stream operators that offer a mechanism to specify when old data is discarded (eviction policy), when to trigger a computation over the accumulated data, what function defines the computation and how it manages intermediate states that lead to the result (which may involve data duplication). Many complex applications such as live machine learning are based on the concept of window [6,8,15]. Active research is carried out both at the level of performance optimizations [6] and expressivity (e.g., extended triggering mechanisms and watermarking [4]).

3 Big Data Use Cases with Emerging Analytics Uncertainty

Big data analytics is data-centric: it needs to constantly adapt and evolve to match the nature of the data available to it. This contrasts the traditional approaches that require the data to adapt to the application. In this section, we develop this perspective using three illustrative use cases where the analytics uncertainty is the main driver for the need to adapt.

A/B Testing: A big data solution running in production often has to solve a difficult trade-off: on one hand it has to be stable enough to deliver consistent results, but on the other hand it needs to be constantly refined to maximize the delivered value (i.e., can further value be extracted?). Despite significant advances in techniques to reduce the cycle of adopting innovation in production through agile DevOps-based software engineering, such approaches are insufficient for big data analytics. As a consequence, *A/B testing* (also called *split-run testing*) is often employed: the main stable solution A is constantly producing results, while an alternative, trial-and-error B solution running in parallel is used to challenge the results. If the B solution obtains a better result, it already replaces the initial result. Eventually, if the B solution consistently delivers better results, then it replaces the A solution and the process restarts.

Machine Learning: A key class of applications that are based on big data analytics is machine learning. Similar with A/B testing, machine learning typically trains and deploys a stable model (i.e. the *champion* model), while at the same time it needs to continuously improve it. To accelerate the improvement of the champion, a typical approach is to construct alternative models (trained with different data, different or variations of the main algorithm, or a combination thereof) that challenge the champion model for the top spot. The challenge in this context is to run both the champion and the challenger model in a seamless fashion, so that they complement each other and generate mutual feedback used for online improvement.

Deep Exploration: Big data is a powerful tool to discover trends and extract insight out of data without necessarily knowing in advance what to look for. The thin line between this and machine learning is given by the goal of the analysis, which for exploration algorithms moves from building a model that solves a particular well defined problem to synthesizing the data, i.e., transforming the data from its raw form into an knowledge augmented form. In this process, various hypothesis and classes of exploration operations are carried with respect to the data [5]. This raises the need to enable efficient multi-path exploration, mainly against large data sets or within performance constrains.

4 Limitations of State-of-Art and Opportunities

Current state of art approaches described in Sect. 2 lack the support to deal with the analytics uncertainty as illustrated in Sect. 3. This happens both at conceptual level (i.e., users lack mechanisms and APIs from the runtime and need to implement their own application-level approach) as well as at runtime level (i.e., the runtime is unaware of the intent of the application to deal with the analytics uncertainty and therefore misses optimization opportunities). In this context, we identify two important dimensions:

The breadth uncertainty: When users need to process the same data set with multiple algorithms simultaneously, they need to manage complex workflows and intermediate states at application level: when to branch into an alternate direction, what intermediate state to compare with, where to roll back to try a different direction, etc. Lacking support from the runtime, users are forced to interact with various layers explicitly (e.g., use the storage layer to persist intermediate states), which can lead to scalability and performance issues. Furthermore, unaware of the relationship between the alternate directions, the runtime will often perform suboptimally (e.g., duplicate data unnecessarily). Thus, an interesting research direction in this context is how to expose a data processing model that natively enables the exploration of alternate directions, which ultimately enables the construction of optimized runtimes.

The temporal uncertainty: In addition to processing the same data set with multiple algorithms simultaneously, an important aspect of the analytics uncertainty is also the temporal dimension. Specifically, as new data is accumulating, the question of "freshness" arises: does it make sense to process all available data or focus only on the most recent data to gain the most relevant insight into the future? Where is the right trade-off, i.e. how far back into the history does it make sense to search? Answering such a question naturally leads to the need of combining batch processing with stream processing. However, the current ecosystem of big data analytics frameworks is fragmented: users have to use different runtimes for different jobs. This becomes an issue both because the application has to explicitly manage different jobs on different runtimes (which leads to complexity) and because the runtimes do not talk to each other (e.g. share common data and states) and as such perform suboptimally. Thus, an interesting research direction is to extend data processing models with native support to focus on the temporal aspect.

5   Conclusions

In this paper we studied the problem of *analytics uncertainty*: how to enable extracting valuable insight from an ever-increasing data market whose usefulness is not known in advance. We identified several use cases where this problem arises and discussed the limitations of existing state of art big data analytics approaches in handling such use cases. Our view argues in favor of the need to build a more expressive data management model that facilitates the construction of corresponding optimized runtimes. In this context, we highlighted the importance of addressing two uncertainty dimensions: *breadth* and *temporal*. Based on the findings, we plan to explore these two directions in future work.

References

1. Flink. https://flink.apache.org/
2. The Zettabyte Era: Trends and Analysis. Cisco Systems, White Paper 1465272001812119 (2016)
3. Akidau, T., Balikov, A., Bekiroglu, K., Chernyak, S., Haberman, J., Lax, R., McVeety, S., Mills, D., Nordstrom, P., Whittle, S.: Millwheel: Fault-tolerant stream processing at internet scale. In: Very Large Data Bases, pp. 734–746 (2013)
4. Akidau, T., Bradshaw, R., Chambers, C., Chernyak, S., Fernndez-Moctezuma, R.J., Lax, R., McVeety, S., Mills, D., Perry, F., Schmidt, E., Whittle, S.: The dataflow model: A practical approach to balancing correctness, latency, and cost in massive-scale, unbounded, out-of-order data processing. Proc. VLDB Endowment 8, 1792–1803 (2015)
5. Cao, L., Wei, M., Yang, D., Rundensteiner, E.A.: Online outlier exploration over large datasets. In: 21th ACM SIGKDD International Conference on Knowledge Discovery and Data Mining, KDD 2015, Sydney, Australia, pp. 89–98 (2015)
6. Carbone, P., Traub, J., Katsifodimos, A., Haridi, S., Markl, V.: Cutty: Aggregate sharing for user-defined windows. In: 25th ACM International on Conference on Information and Knowledge Management, CIKM 2016, pp. 1201–1210 (2016)
7. Dean, J., Ghemawat, S.: Mapreduce: Simplified data processing on large clusters. In: 6th Conference on Symposium on Opearting Systems Design and Implementation, OSDI 2004, pp. 10:1–10:13. USENIX Association, San Francisco (2004)
8. Hammad, M.A., Aref, W.G., Elmagarmid, A.K.: Query processing of multi-way stream window joins. VLDB J. 17(3), 469–488 (2008)
9. Neumeyer, L., Robbins, B., Kesari, A., Nair, A.: S4: Distributed stream computing platform. In: 10th IEEE International Conference on Data Mining Workshops, ICDMW 2010, Los Alamitos, USA, pp. 170–177 (2010)
10. Nicolae, B., Costa, C., Misale, C., Katrinis, K., Park, Y.: Leveraging adaptive I/O to optimize collective data shuffling patterns for big data analytics. IEEE Trans. Parallel Distrib. Syst. (2017)
11. Nicolae, B., Kochut, A., Karve, A.: Towards scalable on-demand collective data access in IaaS clouds: An adaptive collaborative content exchange proposal. J. Parallel Distrib. Comput. 87, 67–79 (2016)
12. Hey, T., Tansley, S., Tolle, K.M.: The Fourth Paradigm: Data-Intensive Scientific Discovery. Microsoft Research, Redmond (2009)
13. Toshniwal, A., et al.: Storm@twitter. In: 2014 ACM SIGMOD International Conference on Management of Data, SIGMOD 2014, Snowbird, USA, pp. 147–156 (2014)
14. Tudoran, R., Costan, A., Nano, O., Santos, I., Soncu, H., Antoniu, G.: Jetstream: Enabling high throughput live event streaming on multi-site clouds. Future Gener. Comput. Syst. 54, 274–291 (2016)
15. Yang, D., Rundensteiner, E.A., Ward, M.O.: Shared execution strategy for neighbor-based pattern mining requests over streaming windows. ACM Trans. Database Syst. 37(1), 5:1–5:44 (2012)
16. Zaharia, M., Chowdhury, M., Das, T., Dave, A., Ma, J., McCauly, M., Franklin, M.J., Shenker, S., Stoica, I.: Resilient distributed datasets: A fault-tolerant abstraction for in-memory cluster computing. In: The 9th USENIX Symposium on Networked Systems Design and Implementation, NSDI 2012, San Jose, USA (2012)
17. Zaharia, M., Das, T., Li, H., Shenker, S., Stoica, I.: Discretized streams: An efficient and fault-tolerant model for stream processing on large clusters. In: 4th USENIX Conference on Hot Topics in Cloud Ccomputing, HotCloud 212 (2012)

Information Extraction and Retrieval

Experiments with Document Retrieval from Small Text Collections Using Latent Semantic Analysis or Term Similarity with Query Coordination and Automatic Relevance Feedback

Colin Layfield, Joel Azzopardi, and Chris Staff[✉]

University of Malta, Tal-Qroqq, Msida 2080, Malta
{colin.layfield,joel.azzopardi,chris.staff}@um.edu.mt

Abstract. Users face the Vocabulary Gap problem when attempting to retrieve relevant textual documents from small databases, especially when there are only a small number of relevant documents, as it is likely that different terms are used in queries and relevant documents to describe the same concept. To enable comparison of results of different approaches to semantic search in small textual databases, the PIKES team constructed an annotated test collection and Gold Standard comprising 35 search queries and 331 articles. We present two different possible solutions. In one, we index an unannotated version of the PIKES collection using Latent Semantic Analysis (LSA) retrieving relevant documents using a combination of query coordination and *automatic* relevance feedback. Although we outperform prior work, this approach is dependent on the underlying collection, and is not necessarily scalable. In the second approach, we use an LSA Model generated by SEMILAR from a Wikipedia dump to generate a Term Similarity Matrix (TSM). Queries are automatically expanded with related terms from the TSM and are submitted to a term-by-document matrix Vector Space Model of the PIKES collection. Coupled with a combination of query coordination and *automatic* relevance feedback we also outperform prior work with this approach. The advantage of the second approach is that it is independent of the underlying document collection.

Keywords: Term similarity matrix · SEMILAR LSA Model · PIKES test collection · Log Entropy

1 Introduction

When databases contain textual documents, retrieving relevant documents can be made more accurate by pre-processing them and making available alternative indexes for search and retrieval. Stankovic *et al.* extract named entities from documents stored in a geological database in Serbian and index both the document

© Springer International Publishing AG 2017
A. Calì et al. (Eds.): IKC 2016, LNCS 10151, pp. 25–36, 2017.
DOI: 10.1007/978-3-319-53640-8_3

content and the named entities to demonstrate that it is superior to using SQL on its own [15]. Successes are also achieved when recognised named entities are linked to 'ground truth' [4, 17] through, for instance, Name Entity Linking (NEL) to Linked Open Data (LOD). In this paper, we describe alternative approaches based on Latent Semantic Analysis [5] and Term Similarity that do not need to recognise named entities, and which achieves greater accuracy on the PIKES test collection [17] than previous work [4].

Corcoglioniti *et al.* first mark up and index the text of documents and then add a number of 'semantic layers' that support Named Entity Linking and represent relations including temporal relations. Queries are similarly processed and documents are retrieved from the collection on the basis of similarity to the query and they are ranked using the semantic information [4]. The advantage of their approach is that they are using external resources to perform Named Entity Linking, which enables them to accurately identify and reason about named entities in and across documents. However, their solution is not scalable (yet).

One of our approaches is based on applying Latent Semantic Analysis (LSA) to the document collection used for retrieval. LSA is known to be computationally expensive given large enough text collections (although approximate techniques based upon ignoring certain word lexical categories, or parts of speech, are achieving good results, e.g., the SEMILAR LSA Models [16]). Our second approach uses one of the SEMILAR LSA Models, which has been produced from an early Spring 2013 English Wikipedia dump, to identify additional terms that are related to terms that appear in queries. This second approach is independent of the PIKES collection and is scalable. Although it does not give results that are as accurate as those obtained by applying LSA directly to the PIKES collection, it still outperforms prior work (see Sect. 4).

The aim of Corcoglioniti *et al.* is to "investigate whether the semantic analysis of the query and the documents, obtained exploiting state-of-the-art Natural Language Processing techniques (e.g., Entity Linking, Frame Detection) and Semantic Web resources (e.g., YAGO, DBpedia), can improve the performances of the *traditional* [our italics] term-based similarity approach" [4]. Our work demonstrates that an approach based on Latent Semantic Analysis outperforms the semantic approach taken by Corcoglioniti *et al.* on the same test collection, as does an approach based on automatic query expansion on a traditional Vector Space Model of the PIKES collection, when, using either approach, *query coordination* is incorporated into the similarity measure and *automatic relevance feedback* is used to re-rank results.

2   Literature Review

There is a significant difference between searching for any document that contains the information you seek in a massive collection like the World Wide Web, and finding all the documents that contain relevant information in small, specialised collections. In the former, many relevant documents are likely to exist, some of which will contain terms that the user expressed in the query. As long as one of those documents is retrieved, the user may be satisfied. In the latter, different terms are likely to be used in documents to represent the concept, so if a predominantly query term-matching retrieval approach is used, only a subset of relevant documents will be retrieved. Of course, queries may be underspecified in either situation, in which case insufficient information is available to disambiguate the query. To address this problem, search results clustering may be used to partition the results list into clusters, where each cluster represents a different query *sense* (e.g., [10,14]). Attempts to improve recall and precision in small document collections include removing, reducing, or otherwise handling ambiguity in documents, and disambiguating query terms or clustering results.

The PIKES document and query test collection was built to showcase that in small collections search is limited because of the vocabulary gap - basically, relevant documents might not contain terms that the user has included in the query, but they might contain related terms [17].

Disambiguating and exposing Named Entities occurring in documents and queries and the relations between the named entities can assist with determining the degree of relevance between a document and a query [4]. Named Entities can be explicitly linked to open data sources and external resources such as WordNet or Wikipedia can be used to determine the degree of similarity between identified relations. Azzopardi and Staff also use Named Entities and relations between them to automatically cluster news reports by news event [2], eventually building a *fused* document containing the information from different reports [1]. However, this is achieved without the utilisation of external resources.

A number of studies show how Information Retrieval (IR) can be enhanced by extracting Named Entities and other semantic information. Stankovic *et al.* extract Named Entities from text files stored in a geological database [15], and index the Named Entities as well as performing full-text indexing on the files. Waitelonis *et al.* proposed marking up documents with semantic information, to help uncover semantic similarity and relatedness, usually by marking up Named Entities with Linked Open Data sources. They produced the PIKES test collection[1]. Their approach "shows that retrieval performance on less than web-scale search systems can be improved by the exploitation of graph information from LOD resources" [17]. Corcoglioniti *et al.* perform full-text indexing on the cleaned PIKES collection, perform Named Entity Linking, and extract relations from the collection, including temporal relations, which are then indexed as separate *semantic layers* [4]. When the collection is queried, the query is similarly processed, relevant documents are retrieved using and the results are ranked using the semantic layers, which usually increases precision.

Rather than using Named Entities or semantic classes/entities, purely statistical techniques can be used to enhance retrieval. One of the original aims of Latent Semantic Analyses (LSA) was to enhance IR by extracting the 'latent semantic structure' between terms and documents [5]. SEMILAR have made

---

[1] The original and cleaned collections are available from http://pikes.fbk.eu/ke4ir.html.

available a number of LSA models built from a Wikipedia dump [16], and identify the Wiki 4 LSA Model to be the best to perform term similarity[2]. To produce these LSA models, they systematically removed lemmas that belong to certain lexical categories. For instance, to produce the Wiki 4 LSA Model, they removed words that do not exist in WordNet[3], adverbs, adjectives, and 'light' verbs from the document collection before indexing the documents and performing Singular Value Decomposition using Latent Semantic Analysis. This produces a term vector matrix for approximately 68000 lemmas. Two term vectors can be compared using cosine similarity, for instance, the result of which indicates the 'relatedness' (or semantic similarity) between the two terms.

3   Our Approach

We begin by giving a general overview of our approach, providing detailed information in the following subsections. We experiment on the same cleaned PIKES test collection as Corcoglioniti *et al.* As described in Subsect. 3.1, we first build a standard Vector Space term-by-document Model of the unannotated PIKES document collection (which we refer to as the PIKES VSM Model). We use the PIKES VSM Model in two distinct approaches.

In the first approach, we derive an LSA Model from the PIKES VSM Model using Latent Semantic Analysis [5]. In the second approach, we use the PIKES VSM Model directly, but we automatically expand queries with related terms before retrieving a list of results. To support query expansion, we transformed an LSA Model independently generated from a Wikipedia dump (Stefanescu *et al.*'s SEMILAR Wiki 4 [16]) into a Term Similarity Matrix (Wiki4TSM).

Each query in the test collection is transformed into a query vector using Log Entropy [6] to calculate the query term weights (Subsects. 3.2 and 3.3) and a ranked list of documents is retrieved from the collection representation using the approach described in Subsects. 3.4 and 3.5. The retrieval method includes *query coordination* and document length normalisation. Inspired by the Lucene search engine approach[4], query coordination is a process through which documents in the results list may have their query-document similarity score modified to take into account the percentage of query terms present in the document.

In our approach, documents in the initial results list are subsequently re-ranked using *automatic* relevance feedback inspired by Rocchio's approach [12]. However, rather than modifying the original query, we re-rank the results in order of weighted similarity to the average vectors derived from the top-n documents in the current results list (see Subsect. 3.5). Typically, only the top-one or top-two documents in the results list are used, because the PIKES collection contains only 331 documents with between one and twelve relevant documents per query according to the Gold Standard[5]. The top-n documents used to derive

---

[2] Wiki 4 and other LSA Models are available from http://www.semanticsimilarity.org.

[3] http://wordnet.princeton.edu.

[4] https://www.elastic.co/guide/en/elasticsearch/guide/current/
practical-scoring-function.html.

[5] http://pikes.fbk.eu/ke4ir.html.

the average document vector keep their original ranking following automatic relevance feedback.

3.1 Generating the Document Collection Representations

To produce the PIKES VSM Model used in both our approaches, stop words[6] are removed from documents, the remaining terms are stemmed using the Porter Algorithm [11], and term weights are calculated using Log Entropy [6].

In the first approach, we generate an LSA model from the PIKES VSM Model, which we refer to as the PIKES LSA Model. We decompose the PIKES VSM Model using LSA with 200-dimensions (determined empirically). In the second approach we use the PIKES VSM Model directly, with an optional preprocessing step to find related terms for each query term using a Term Similarity Matrix derived independently of the PIKES document collection, (see Subsect. 3.2). The expanded query is submitted to the PIKES VSM Model.

The motivation for trying these two approaches is to overcome the vocabulary gap present in small document collections. Given that the collection may contain documents relevant to the query but which do not contain the query terms, we compare the approaches of (i) using 'latent' semantics within the collection to expose terms that are related to each other and which are present in the collection representation itself (i.e., using the PIKES LSA Model); and, (ii) using a term similarity matrix derived from an independent collection (i.e., our Wiki4TSM is derived from SEMILAR's Wiki 4 LSA Model) to expand the query to include terms that are related to the query terms.

3.2 Finding Terms Related to Query Terms

The SEMILAR Wiki 4 LSA Model is a matrix of lemma vectors that can be used to calculate the semantic similarity or 'relatedness' between any two lemmas. Wiki 4 contains more than 68000 unique lemmas. We are unable to fold the PIKES document collection and queries into the Wiki 4 LSA Model, because the publicly available model does not support this. Instead, we process the Wiki 4 LSA Model matrix to generate a lemma-to-lemma similarity matrix consisting of the cosine similarity scores between each lemma pair in the Wiki 4 vocabulary (inspired by the approach reported in [8]). This yields the Wiki4TSM Term Similarity Matrix where for a lemmatized term the intersection of the lemma and any other lemma is the 'relatedness' between the lemmas, represented as a similarity score. Also, the elements in a lemma's vector of similarity scores can be ordered according to lemma similarity to obtain a ranked list of similar or related lemmas. As Wiki4TSM is derived from an independent source, it is possible, for any document collection, that there are lemmas in the collection (e.g., PIKES VSM Model) that are not present in Wiki4TSM and *vice versa*. Indeed, 6396 of 16677 unique lemmas in PIKES (38.35%) are missing from Wiki4TSM. However, these

---

[6] The stop word list is available at http://www.lextek.com/manuals/onix/stopwords1.html.

missing terms are generally non-English words, 'light' verbs, adverbs, and adjectives excluded by the SEMILAR team in the construction of the Wiki 4 LSA Model, misspelt words, hyphenated words, numerals, proper nouns (therefore including Named Entities), words that contain numerics, acronyms, abbreviations, and other terms that are not found in WordNet (which were excluded by the SEMILAR team). In our second approach, we expand the query vector with related terms obtained from the Wiki4TSM.

3.3   Query Processing

The PIKES test collection contains 35 queries. How we process the query depends on whether we are using the PIKES LSA Model, or whether we expand the query using the Wiki4TSM and submit it to the PIKES VSM Model. In either case, stop words are removed from the query before it is submitted.

To find related terms in the Wiki4TSM, query terms are lemmatized using the Stanford Core Lemmatizer[7]. As explained in Subsect. 3.2, the Wiki 4 LSA Model constructed by the SEMILAR team has a vocabulary of lemmas. For each unique lemma in the query, we extract from Wiki4TSM a similarity vector where the elements represent the similarity scores of that lemma to other lemmas in the vocabulary. We then average the similarity vectors extracted for each lemma in the query - this results in a new vector that represents the average similarity of the query as a whole to the other lemmas that are not present in the query. From this average similarity vector, we remove those lemmas whose similarity is less than 0.7, and extract the top 5-weighted lemmas from the remaining ones (both determined empirically). For example, given the original query 'bridge construction' we identify and add to the query vector terms that are related to both 'bridge' and 'construction', if any. The related lemmas we extract from the Wiki4TSM are 'construct' 'girder' 'abutment', 'truss', and 'span'. As the identification of related lemmas is totally independent from the PIKES document collection, as mentioned in Subsect. 3.2, there is no guarantee that the related lemmas are present in the collection, or that query terms have a vector for the lemmatized term in Wiki4TSM.

A query submitted to the PIKES VSM Model may contain additional related lemmas after being expanded. The lemmas in the query are stemmed using the Porter Stemmer and the unique stems are weighted using Log Entropy [6]. The term weight for stems added to the query vector as related terms is dampened by a factor of 0.5 (determined empirically).

3.4   Query Coordination and Document Length Normalisation

Queries are becoming increasingly verbose [7]. Rather than containing specific query terms, queries are becoming more natural language-like. Users may require assistance to construct a query, or wish, based upon initial results, to modify a query by increasing or reducing the importance of one or more terms.

---

[7] http://stanfordnlp.github.io/CoreNLP/.

In *query coordination*, users can be assisted to construct a search query [13]. Lucene's query coordination approach is automatic. Lucene retrieves documents that contain any of the search terms, 'rewarding' each document based upon the percentage of query terms the document contains[8]. In our approach, the query vector derived in the previous step (Subsect. 3.3) is submitted to the document collection representation (the PIKES LSA Model or the PIKES VSM Model, depending on the approach used, bearing in mind that the LSA Model is the VSM Model with reduced dimensionality). The scoring function given in Eq. 1, inspired by Lucene's Practical Scoring Function[9], is used to retrieve a ranked list of relevant documents, together with their similarity score $score(q, d_i)$. The query coordination score is represented as the percentage of lemmas in the *original* query that are present in the document ($coord(q, d_i)$). The scoring function adjusts the similarity scores to take document length normalisation into account (see Eq. 1). We have also experimented with performing query coordination using the related lemmas that may have been added to the query vector in the Wiki4TSM approach (see Eq. 2).

$$score(q, d) = (\boldsymbol{q} \cdot \boldsymbol{d}) \times coord(q, d) \times docLenNorm(d) \tag{1}$$

$$rscore(q, d) = score(q, d) + (0.5 \times score(q, d) \times relQueryCoord(q, d)) \tag{2}$$

where:

$$coord(q, d) = \frac{|q \cap d|}{|q|} \qquad docLenNorm(d) = \frac{1}{\sqrt{\|d\|}}$$

$$relQueryCoord(q, d) = \frac{|rel(q) \cap d|}{|rel(q)|}$$

q = Set of terms in query

d = Set of terms in document

$\boldsymbol{q}$ = Vector space representation of q

$\boldsymbol{d}$ = Vector space representation of d

$|q|$ = Cardinality of the q (number of different terms in q).

$\|d\|$ = Length of d (total number of terms in d).

$rel(q)$ = Set of related terms to q

3.5   Pseudo-Relevance Feedback and *Automatic* Relevance Feedback

Pseudo-Relevance Feedback and Query Expansion (e.g., [3,9]) can be used to re-rank a ranked list of retrieved results. Such approaches assume that the

---

[8] Concisely explained at https://www.elastic.co/guide/en/elasticsearch/guide/current/practical-scoring-function.html.

[9] https://lucene.apache.org/core/4_6_0/core/org/apache/lucene/search/similarities/TFIDFSimilarity.html.

top-n retrieved documents are more relevant to the query than the bottom-m documents. The initial query is automatically modified to add high weighted terms present in the top-n documents that are missing from the query (similarly, removing terms from the query that are present only in the bottom-m retrieved ranked documents). The documents ranked $n + 1$ to $m - 1$ are re-ranked using the reformulated query. Mitra's approach is to modify the query assuming that of 1,000 retrieved results, the top-20 are relevant and the bottom-500 are non-relevant [9]. Automatic Query Expansion or Reformulation has also been used on structured data. Yao *et al.*'s approach [18] requires the collection to be pre-processed to identify terms in the database that can substitute or enhance terms in the query. A *"term augmented tuple graph"* (TAT) is used to model term relationships between terms (words and phrases) automatically extracted from the structured database. Random walks through the TAT are used to determine the probability of textual similarity.

Our approach to document re-ranking is motivated by Mitra's [9]. However, for small text collections, we experimented with *automatic* relevance feedback to re-rank documents in the results list with respect to the top-n documents only. In these experiments, we generate the average document vector representing just the top-1 or top-2 documents in the results list. Then the standard Cosine Similarity Measure is applied to determine the similarity score between this average vector and the other document vectors in the results list. This score is dampened by a factor of 0.7 (determined empirically) and the result is added to the document's original query-document similarity score. This step can lead to document re-ranking, but the top-n documents used to generate the average top-ranked documents vector keep their original rank.

4   Evaluation

The PIKES collection "consists of 331 articles from the yovisto blog on history in science, tech, and art. The articles have an average length of 570 words, containing 3 to 255 annotations (average 83) and have been manually annotated with DBpedia entities" [17]. We use a version cleaned by the KE4IR team that does not include the annotations[10] to generate the PIKES VSM Model and the PIKES LSA Model.

We compare the results of nine runs. The runs with their settings are given in Table 1. Runs 1–6 are performed on the PIKES LSA Model. There is one basic baseline run (Run 1 with no query coordination and no relevance feedback); two runs for relevance feedback only (using either the top-1 or top-2 documents in Runs 2 and 3 respectively); one run with query coordination only (Run 4); and two runs with relevance feedback and query coordination combined (again, using either the top-1 or top-2 documents for relevance feedback in Runs 5 and 6 respectively). Runs 7–9 are performed on the PIKES VSM Model with automatic relevance feedback and query coordination of the terms in the initial query. Run 7 does not expand the query. Runs 8 and 9 automatically expand

---

[10] Available from http://pikes.fbk.eu/ke4ir.html.

Table 1. Our runs and their description

	Run 1	Run 2	Run 3	Run 4	Run 5	Run 6	Run 7	Run 8	Run 9
PIKES LSA Model	✓	✓	✓	✓	✓	✓			
PIKES VSM Model							✓	✓	✓
Wiki4TSM								✓	✓
Query expansion								✓	✓
Relevance feedback		top-1	top-2		top-1	top-2	top-1	top-1	top-1
Query coordination original terms				✓	✓	✓	✓	✓	✓
Query coordination related terms									✓

the queries with related terms obtained from Wiki4TSM. Run 9 also performs query coordination on the related terms added by query expansion.

The results of our experiments are given in Table 2 using the same measures used by Corcoglioniti *et al.* for KE4IR [4]. The Textual results are the best of the baselines used by the KE4IR team to evaluate their approach. The measures are Precision at Rank n (P@1, P@5, P@10), Normalised Discounted Cumulative Gain (NDCG) and NDCG@10, Mean Average Precision (MAP) and MAP@10.

Table 2. Our results, compared to Textual and KE4IR

	P@1	P@5	P@10	NDCG@10	NDCG	MAP@10	MAP
Textual	0.9430	0.6690	0.4530	0.7820	0.8320	0.6810	0.7330
KE4IR	0.9710	0.6800	0.4740	0.8060	0.8540	0.7130	0.7580
Run 1	0.9714	0.6686	0.4571	0.7911	0.8427	0.6967	0.7505
Run 2	0.9714	0.6629	0.4714	0.7955	0.8478	0.7052	0.7573
Run 3	0.9714	0.6800	0.4714	0.7982	0.8502	0.7131	0.7631
Run 4	**1.0000**	0.6743	0.4765	0.8034	0.8452	0.7090	0.7568
Run 5	**1.0000**	0.6686	**0.4886**	**0.8145**	**0.8655**	0.7217	**0.7809**
Run 6	**1.0000**	**0.6914**	**0.4886**	0.8103	0.8604	**0.7230**	0.7795
Run 7	**1.0000**	0.6686	**0.4829**	0.8100	0.8621	0.7153	0.7736
Run 8	**1.0000**	0.6857	**0.4829**	0.8108	**0.8629**	0.7217	**0.7800**
Run 9	**1.0000**	0.6857	**0.4829**	0.8071	0.8593	0.7194	0.7777

In the experiments performed on the PIKES LSA Model our results only outperform KE4IR on all measures in Run 6. This indicates that query coordination and automatic relevance feedback individually are insufficient to outperform KE4IR, though query coordination on its own (Run 4) beats our baseline (Run 1) on all measures. Run 6 combines them, uses the top-2 documents for automatic relevance feedback and outperforms KE4IR on all measures.

When we use the PIKES VSM Model together with query coordination and relevance feedback but without query expansion (Run 7), we outperform KE4IR on most but not all measures. When we add automatic query expansion using Wiki4TSM (Runs 8 and 9) we outperform KE4IR on all measures. However, Run 9 shows that performing query coordination on the related terms as well harms ranking accuracy and does not improve precision compared to excluding the related terms from the query coordination step (Run 8). Indeed, it may even counteract the benefits of performing query expansion at all, as the ranking accuracy results (NDCG and NDCG@10) are worse than those achieved by Run 7 (PIKES VSM Model without query expansion)[11].

5   Discussion

We have shown that generating an LSA Model of a document collection or using an independently generated Term Similarity Matrix to automatically expand a query with related terms for use with a Vector Space representation of the same collection, both coupled with query coordination and *automatic* relevance feedback (Subsect. 3.5), outperform current approaches that perform Named Entity Linking and relationship representation to support semantic search on a small document collection. Our best performing runs on the PIKES LSA Model (Runs 5 and 6) differ only on whether the top-1 or top-2 documents in the results list are used for automatic relevance feedback to re-rank the results. However, Run 5 fares badly on the P@5 measure, failing to outperform KE4IR and Textual, and beating only Run 2 (Run 2 is identical to Run 5 except that it does not perform query coordination).

The main difference between the PIKES LSA Model and PIKES VSM Model approaches is that in the former related terms are automatically discovered from within the collection itself through the process of decomposition of the VSM Model resulting in an implicit representation of the degree of relatedness between terms. Using the PIKES VSM Model with query coordination and automatic relevance feedback is usually, but not always, sufficient to outperform prior work (compare Run 7 to KE4IR's results in Table 2), but additionally performing query expansion to include in the query vector related terms retrieved from an independently generated Term Similarity Matrix always outperforms prior work (Runs 8 and 9 compared to KE4IR's results).

---

[11] https://www.elastic.co/guide/en/elasticsearch/guide/current/
practical-scoring-function.html explains that query coordination may not be effective when the query contains synonyms.

6 Conclusions

Small document collections generally contain documents that may be semantically relevant to a query but that may contain terms that are different from those expressed in the user query. Although this is also likely in large document collections, it is generally less of a problem unless the users needs to obtain all, rather than just some, relevant documents. In small collections, the vocabulary gap may result in too few or no relevant documents being retrieved.

Two approaches to indexing and retrieving textual documents were presented. Both yield greater accuracy than prior work on the same test collection. Stop words were removed from the collection and the remaining unigrams were stemmed and indexed as a Vector Space term-by-document matrix (PIKES VSM Model). In the first approach, related terms in the PIKES collection are automatically discovered and represented using LSA (PIKES LSA Model). In the second approach, the query is automatically expanded with related terms from the Term Similarity Matrix (derived from an LSA model of a Wikipedia dump). An adaptation of Lucene's Practical Scoring Function is used to retrieve ranked documents from the PIKES VSM Model, using query coordination to reward documents based on the percentage of query terms they contain. *Automatic* relevance feedback is used to re-rank documents in the results list.

In the PIKES LSA Model approach, the 'latent' semantics exposed in the collection is highly dependent on the documents in the collection. In the second approach, an LSA Model generated from Wikipedia is used to build a collection-independent Term Similarity Matrix that is then used to identify and expand the query with related terms. Both approaches outperform prior work when evaluated using the same PIKES test collection. Prior work had identified mentioned named entities, linked them to their open data, and identified and explicitly represented semantic relationships between named entities mentioned in the same document. Corcoglioniti *et al.* observed that their approach is not scalable [4]. Our first approach, deriving an LSA model of the text collection directly, may also prove difficult to scale given a large enough text collection (this is a known limitation of LSA). However, in our second approach, we utilised an existing, independently derived LSA model to perform automatic query expansion, which, although not as accurate as our first approach, still outperforms prior work.

References

1. Azzopardi, J., Staff, C.: Fusion of news reports using surface-based methods. In: WAINA 2012: Proceedings of the 26th International Conference on Advanced Information Networking and Applications. Workshops, pp. 809–814. IEEE Computer Society, Los Alamitos (2012)
2. Azzopardi, J., Staff, C.: Incremental clustering of news reports. Algorithms **5**(3), 364–378 (2012)
3. Carpineto, C., Romano, G.: A survey of automatic query expansion in information retrieval. ACM Comput. Surv. **44**(1), 1:1–1:50 (2012)

4. Corcoglioniti, F., Dragoni, M., Rospocher, M., Aprosio, A.P.: Knowledge extraction for information retrieval. In: Sack, H., Blomqvist, E., d'Aquin, M., Ghidini, C., Ponzetto, S.P., Lange, C. (eds.) ESWC 2016. LNCS, vol. 9678, pp. 317–333. Springer, Heidelberg (2016). doi:10.1007/978-3-319-34129-3_20

5. Deerwester, S., Dumais, S.T., Furnas, G.W., Landauer, T.K., Harshman, R.: Indexing by latent semantic analysis. J. Am. Soc. Inf. Sci. **41**(6), 391–407 (1990)

6. Dumais, S.: Improving the retrieval of information from external sources. Behav. Res. Methods Instrum. Comput. **23**(2), 229–236 (1991)

7. Huston, S., Bruce Croft, W.: Evaluating verbose query processing techniques. In: Proceedings of the 33rd International ACM SIGIR Conference on Research and Development in Information Retrieval, SIGIR 2010, pp. 291–298. ACM, New York (2010)

8. Jorge-Botana, G., Olmos, R., Barroso, A.: The construction-integration framework: a means to diminish bias in LSA-based call routing. I. J. Speech Technol. **15**(2), 151–164 (2012)

9. Mitra, M., Singhal, A., Buckley, C.: Improving automatic query expansion. In: Proceedings of the 21st Annual International ACM SIGIR Conference on Research and Development in Information Retrieval, SIGIR 1998, pp. 206–214. ACM, New York (1998)

10. Navigli, R., Vannella, D.: SemEval-2013 task 11: word sense induction and disambiguation within an end-user application. In: Second Joint Conference on Lexical and Computational Semantics (*SEM), Proceedings of the Seventh International Workshop on Semantic Evaluation (SemEval 2013), vol. 2, pp. 193–201. Association for Computational Linguistics, Atlanta, June 2013

11. Porter, M.F.: An algorithm for suffix stripping. Program **14**(3), 130–137 (1980)

12. Rocchio, J.J.: Relevance feedback in information retrieval. In: Salton, G. (ed.) The SMART Retrieval System: Experiments in Automatic Document Processing, pp. 313–323. Prentice-Hall, Englewood Cliffs (1971)

13. Spoerri, A.: How visual query tools can support users searching the internet. In: 2014 18th International Conference on Information Visualisation, pp. 329–334 (2004)

14. Staff, C., Azzopardi, J., Layfield, C., Mercieca, D.: Search results clustering without external resources. In: Spies, M., Wagner, R.R., Min Tjoa, A. (eds.) Proceedings of the 26th International Workshop on Database and Expert Systems Applications DEXA 2015, Valencia, Spain, 1–4 September 2015, pp. 276–280 (2015)

15. Stanković, R., Krstev, C., Obradović, I., Kitanović, O.: Indexing of textual databases based on lexical resources: a case study for Serbian. In: Cardoso, J., Guerra, F., Houben, G.-J., Pinto, A.M., Velegrakis, Y. (eds.) KEYSTONE 2015. LNCS, vol. 9398, pp. 167–181. Springer, Heidelberg (2015). doi:10.1007/978-3-319-27932-9_15

16. Stefanescu, D., Banjade, R., Rus, V.: Latent semantic analysis models on wikipedia and tasa. In: Calzolari, N., Choukri, K., Declerck, T., Loftsson, H., Maegaard, B., Mariani, J., Moreno, A., Odijk, J., Piperidis, S. (eds.) Proceedings of the Ninth International Conference on Language Resources and Evaluation (LREC 2014). European Language Resources Association (ELRA), Reykjavik, May 2014

17. Waitelonis, J., Exeler, C., Sack, H.: Linked data enabled generalized vector space model to improve document retrieval. In: Proceedings of NLP and DBpedia 2015 Workshop in Conjunction with 14th International Semantic Web Conference (ISWC 2015), CEUR Workshop Proceedings (2015)

18. Yao, J., Cui, B., Hua, L., Huang, Y.: Keyword query reformulation on structured data. In: 2012 IEEE 28th International Conference on Data Engineering (ICDE), pp. 953–964. IEEE (2012)

Unsupervised Extraction
of Conceptual Keyphrases from Abstracts

Philipp Ludwig, Marcus Thiel[✉], and Andreas Nürnberger

Faculty of Computer Science, Otto von Guericke University Magdeburg,
Magdeburg, Germany
marcus.thiel@ovgu.de

Abstract. The extraction of meaningful keyphrases is important for
a variety of applications, such as recommender systems, solutions for
browsing of literature, or automatic categorization of documents. Since
this task is not trivial, a great amount of different approaches have been
introduced in the past, either focusing on single aspects of the process
or utilizing the characteristics of a certain type of document. Especially
when it comes to supporting the user in grasping the topics of a document (i.e. in the display of search results), precise keyphrases can be very
helpful. However, in such situations usually only the abstract or a short
excerpt is available, which most approaches do not acknowledge. Methods based on the frequency of words are not appropriate in this case, since
the short texts do not contain sufficient word statistics for a frequency
analysis. Secondly, many existing methods are supervised and therefore depend on domain knowledge or manually annotated data, which
is in many scenarios not available. Therefore we present an unsupervised graph-based approach for extracting meaningful keyphrases from
abstracts of scientific articles. We show that even though our method is
not based on manually annotated data or corpora, it works surprisingly
well.

1 Introduction and Motivation

In the past years, the size of online available document collections has been
steadily increasing and users face a larger number of potential interesting documents each day. Browsing these collections, while necessary e.g. for researchers
who are moving into a new domain, is very time-consuming and cost-intensive.
Therefore, various approaches have been suggested in the past to aid readers in
not getting overwhelmed with all the potential relevant material and to guide
them in navigating through huge collections.

Since manually navigating through the collection is not possible due to the
sheer amount of items, common solutions include search engines, which present
the user a list of documents based on an entered query, or systems allowing the
user to easily navigate through the material [5]. In this context, the extraction
of *conceptual keyphrases* plays a vital role for performing the described sorting.
Detecting the most fitting keyphrases is a necessary step, as it supports the user

© Springer International Publishing AG 2017
A. Calì et al. (Eds.): IKC 2016, LNCS 10151, pp. 37–48, 2017.
DOI: 10.1007/978-3-319-53640-8_4

in grasping the topics of the search results without the need of examining each item.

Definition of the Term *Conceptual Keyphrase*
Before going further into detail, we need to define the term *conceptual keyphrase*. In the past, the expression *keyphrase* has been used by various authors in different ways, sometimes as a synonym for the term *keyword*. However, since phrases are more suitable for describing documents, as for example shown by Popova and Khodyrev [12], who evaluated the quality of keyphrases compared to single words (keywords), we define a *keyphrase* as a combination of one or more words, where most or all of the words are keywords, and some of them may be stop-words. An example for a keyphrase is the expression "connectivity in wireless mesh networks", where the words "connectivity", "wireless", "mesh" and "network" can be considered keywords, while the word "of" can be considered a stop-word. Based on this definition, we call a keyphrase a *conceptual keyphrase* if it is suitable for describing the topic or a sub-topic of a document. Ideally, a conceptual keyphrase can be directly used as a conceptualization of the information space.

In this paper, we propose an improved method for extracting conceptual keyphrases from abstracts of single documents, without relying on domain knowledge or manual annotation. We show that our approach, while easy to implement, is sufficient to grasp the topics of a variety of documents. The remainder of this document is structured as follows: In the following section we describe related work with focus on keyphrase extraction. Section 3 outlines our graph-based approach and how we extract conceptual keyphrases from abstracts. The design of our evaluation and the accompanying outcomes are presented in Sect. 4. Finally, we conclude our work in Sect. 5, summarizing the results and providing ideas for further research on this topic.

2   Related Work

In the last few years there has been a growing interest on ideas and implementations for keyphrase extraction. The literature shows a variety of methods on the extraction of keywords or keyphrases, which can be categorized in classification based methods (supervised) or ranking based methods (unsupervised), as shown by Lahiri et al. [10]. In this work we focus on unsupervised methods, because the need of manually pre-labeled training data for the supervised approaches comes with additional costs, since the expertise of domain experts is needed when processing i.e. technical documents. Due to this reason, these methods are usually not appropriate for processing large document collections, since this would require experts from several fields.

Unsupervised Keyphrase Extraction
In the area of unsupervised keyphrase ranking methods, textrank and pagerank are among the ones which are used most frequently. Beside these, various other graph-based variants have been proposed; a comprehensive overview has

been presented by Slobodan et al. [3]. Sadly, a lot of them suffer from several limitations, such as utilizing time-intensive algorithms which are not feasible for quickly enhancing the results of a search engine or basing their ranking methods on the word frequency, which is unsuitable for extracting keyphrases from abstracts, since the words in these types of texts appear at maximum two or three times. Even if we ignore this limitation, a simple approach like this can easily neglect potential candidates which might have been a good choice to represent the document, but whose appearance is not as common as other ones. The authors Lossio-Ventura et al. try to overcome this problem by using a different ranking method which examines the neighbors of a term in the document's word graph and uses linguistic pattern probability to prefer words which are important in the specific domain [11]. While they show that their approach is sufficient for their use case, it suffers from the limitation of the need for a homogeneous document corpus of the target domain. Therefore, it is not suitable in our case of enriching search results in a (possible cross-domain) search engine.

Utilizing a Word Graph

Even though manually annotated data or corpora are not always available, a graph built upon the words of the input text can still be utilized to evaluate the content of the text. This approach has been proven useful by various authors, especially since the centrality measures can provide information about the potential words in candidate keywords and keyphrases, as Lahiri et al. have shown [10].

When it comes to using document abstracts, Xie used a graph to select noun phrases suitable for automatic text summarization [18], supported by a decision tree. The author evaluated two cases: Building one graph for the complete abstract and building multiple sub-graphs, i.e. one for each sentence. The findings suggest that when using multiple sub-graphs, the degree measure outperforms the more complex centrality measures, which is not surprising as one sentence hardly contains enough edges for calculating e.g. the betweenness centrality of a node.

Šišović et al. focused on extracting keyphrases from web documents using a graph-based method [16]. They built the graph based on the whole document collection (in contrast to graphs built on single documents), whereas a vertex represents a word and the edges represent their co-occurrence (in a sentence), and selected the top vertices based on the selectivity measure, which they defined as the relationship between a vertex' degree and weight. Their results indicate that this approach is suitable for dealing with multi-topic documents.

Using all available documents for keyphrase extraction in comparison to focusing on single documents has several advantages, such as the possibility to generate domain specific stop words [13,14]. This approach can be taken further by including other external sources such as (technical) dictionaries or social web platforms, as demonstrated by Barla and Bieliková [2]. Using a set of documents of course requires that the articles have been at least roughly manually pre-selected to make sure that only documents of the same domain appear in the collection. Therefore, these kind of approaches come with additional effort

and are not practical in situations where a lot of new material is added in a
short amount of time and supposed to be analyzed on-the-fly, as for example at
on-line libraries. For these reasons, we focus on techniques for single document
keyphrase extraction in the remainder of this document.

Another aspect to consider is the inclusion of stop-words into the keyphrases,
such as in "state of the art". While in most of the previous work keyphrase candidates are usually only selected by choosing all words between two stop-words,
Rose et al. presented a method which supports generating adjacent keyphrases
in a multi-iterative process [14]. From this set of possible candidates, they select
the best keyphrases based on a scoring method, which operates on the frequency
of the words.

Wang et al. argue that a common problem of the known keyphrase extraction methods is that they only can extract words which are contained in the
source document, even though for certain articles more abstract words which
define the actual topic of the text may be more appropriate [17]. Therefore,
they added available datasets of *embedded word vectors* to enhance the candidate keyphrases. The authors show that their method is suitable for generating
additional keywords and they suggest that a domain-specific previously trained
word vector dataset might improve the result further, however, their approach
needs a previously trained word embedding set, which is expensive to compute
as the authors admit themselves.

In summary, a lot of the proposed methods for keyphrase extraction utilize a word graph, which seems reasonable especially when abstracts are to be
processed. However, most of them rely on external sources, such as additional
documents from the same domain, manually trained data or similar material,
which is expensive in terms of time and costs. Therefore, an analysis of using
a word graph without prior domain knowledge or the application of a corpus is
sensible, as this enables the use of this method in cases where domain knowledge
or a corpus is not available.

3   Creating Conceptual Keyphrases from Abstracts

The procedure of creating conceptual keyphrases takes part in two steps: First,
we extract all possible candidates from the input text. Then, we evaluate their
quality using our ranking method and select the best conceptual keyphrases.

For the candidates that we can get from the abstract, a simple n-gram based
approach is feasible. Such n-grams have to be constrained in order to create
usable candidates. One of these constraints is the sentence boundary, since we
assume that there are no keyphrases that cross these sentences boundaries. Generating keyphrase candidates without acknowledging the sentence boundaries is
possible, though this approach very likely results in more junk phrases.

Another constraint are stop-words: Stop-words are language-dependent,
therefore a prior language identification [8] of the abstract may be necessary.
We follow the idea of Rose et al. and select only those conceptual keyphrase
candidates, that neither begin nor end with a stop-word [14]. Stop-words inside

the keyphrase are permissible, like in "connectivity in wireless networks". Even though this may throw out feasible candidates for keyphrases, depending on the stop-word list used, this will still provide a good junk filter and is therefore endorsed.

Since keyphrases can consist of a variable number of words, all possible combinations of n-grams that fulfill the mentioned constraints need to be created from the input text. For example, this would result in not only the creation of the keyphrase "Statistical Learning Theory", but also the subphrases "Statistical Learning", "Learning Theory" and all the individual words. Due to that, the keyphrase candidate set will be quite large even for a short abstract. It is therefore necessary to purge the set of non-informative phrases and rank them afterwards. In order to do that, we need to employ a scoring of the conceptual keyphrase candidates, which we base on the analysis of the word graph.

3.1 Building a Word Graph

As we discussed in Sect. 2, for the task of appraising the extracted keyphrases a graph-based approach is most suitable since we can utilize various graph measures to judge the usefulness of certain words and don't need to depend on the word frequency. When building such a graph, a variety of different approaches is possible. The final version of the algorithm is being presented in pseudo code in Algorithm 1 and will be explained in the following. As described above, we split the input text into sentences and create edges in respect to the words in each sentence, parsing them to iteratively build the graph.

Algorithm 1. Building the word graph

1: **function** BUILDGRAPH(*text*)
2:      *sentences* ← *extractSentences(text)*
3:      **for** *sentence* in *sentences* **do**
4:          *words* ← *tokenize(sentence)*
5:          *words* ← *lowercaseAndStemWords(words)*
6:          *words* ← *removeStopwords(words)*   ▷ stopwords are stemmed
7:          *prevWord* ← ∅
8:          **for** *word* in *words* **do**
9:              **if** *word* is not ∅ **then**
10:                 *addEdge(prevWord, word)*   ▷ Increments weight, if edge exists
11:                 *prevWord* ← *word*

In this regard, a decision has to be made regarding the connectivity of the words, as shown in Fig. 1. In the first variant, every word in the sentence is connected with each other. This strategy has the advantage of possibly identifying conceptual keyphrase candidates such as "network connectivity" in "connectivity in wired and wireless networks", however our preliminary experiments suggested that this approach generates to much junk keyphrases which do not make any

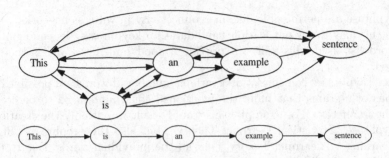

Fig. 1. Two possibilities for building a word graph for one sentence.

sense in the English language. Therefore, we retain the second approach of connecting only neighboring words with each other; not only has this method the upside of only suggesting conceptual keyphrase candidates which are contained in the source material, additionally the resulting graph is much smaller in terms of the number of edges which results in shorter computation times. Since we are concerned with the actual connections and their directions, we limit the graph to a directed variant.

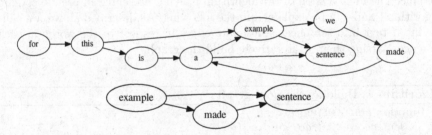

Fig. 2. Graph with and without stopwords, built from the sentences "This is an example sentence" and "For this example, we made a sentence".

In addition to the matter of the word connectivity, there are two options on how to create the nodes and edges of the graph based on the words taken. The first and most straightforward method concerns itself with building the word graph as-is, meaning we take all words in the sentence and connect the adjacent ones. This usually results in a graph where different keyphrases have a connection on a stop-word. Since these kind of connections do not have much semantics, it is feasible to create a graph without all the stop-words. Therefore we bridge the stop-words in the graph and leave them out completely. Figure 2 shows an example of both variants for two sentences: "This is an example sentence" and "For this example, we made a sentence". Leaving out the stop-words did not only make the graph much smaller, but also removed all the semantic useless connections. Therefore it could be argued that the latter option creates a graph with semantic connections, whereas the former concerns itself with syntactic edges.

In order to enrich the graph structure with statistical value, we also add weights to the connecting edges of the graph. A weight represents the co-occurrence of two words in the text. Since this is a directed graph, the co-occurrence is limited to the sequence of the words. In order to decrease the sparseness of the graph (resulting due to the length of abstracts in general), we consider to stem the keywords before adding them as nodes. Adding "Learning" and "Learn" to the graph would therefore account for the same node. This may result (based on the quality of the used stemming approach) in erroneous semantic grouping of the words, but increases the possibility of word repetitions, consequently resulting in better statistical values. For our approach, we applied a simple Porter stemmer. Additionally, the word stems are added in lower case only.

3.2 Scoring Keyphrase Candidates

After building the graph, we need to define how we evaluate the meaningfulness of the single words which are determined to be part of the future conceptual keyphrase candidates. As we discussed in Sect. 2, the degree of a vertex is not useful when dealing with networks comprised of document abstracts, therefore we turn our attention to centrality measures. These kind of measures can be used to determine what kind of role a vertex plays in a network, as defined by Freeman [6]. From all the different centrality measures to pick, which can be used for keyphrase extraction [10], we decided to use the *betweenness centrality* and *closeness centrality* for reasons which are outlined as follows.

By definition, the betweenness centrality indicates how much of the shortest paths between all pairs of vertices in the network go through a given vertex. Therefore in our scenario, the betweenness centrality can be seen as a measure on how much influence a word has in the text. A word with a high betweenness centrality is considered to bridge otherwise unconnected words. We expect these words to be underlying technologies, ideas or methods used, that connect the problem and the solution for it presented in the paper (abstract). In our implementation, we utilized the algorithm presented by Brandes [4] for the calculation of the betweenness centrality.

The closeness centrality is defined as the reciprocal of the sum of minimal distances from a given vertex to all other vertices. In terms of keyphrase evaluation, we can see the closeness centrality of a word as a magnitude of how reachable other words in the graph are. Therefore a keyphrase with words having a high closeness centrality is expected to deal with a wider range of topics or concerning itself with a very central topic of the abstract. In our implementation, we calculated the closeness centrality using the algorithm presented by Sariyüce et al. [15].

In order to generate closeness and betweenness values for keyphrases made out of several keywords, we simply sum up the corresponding values of the underlying words in the graph. Stop-words are not included in the graph and are therefore not considered here as well. Simply summing the values up usually results in preferring longer keyphrases, which is our case favorable, since those also tend to better convey concepts of a paper (abstract). As an additional score we consider the average over all out-edge weights for keyphrases containing only a single word, and the average of the connecting edge weights of the path of keywords in longer keyphrases. The former can be interpreted as the connectedness of the single keyword to the graph and the latter as the connectedness of the keywords inside the keyphrase.

3.3 Purging Non-informative Keyphrase Candidates

In most keyphrase extraction algorithms, keyphrases with small frequencies are removed from the candidate set. Removing all keyphrases with a frequency of 1 would already reduce the size of the candidate set considerably. Due to the generally unsufficient amount of text in abstracts and therefore repetitions of phrases, a lot of "good" candidates only appear once and therefore would be filtered out by this approach. This results in the need of other filtering methods not based on word statistics.

The easiest and most straightforward idea is to remove all keyphrase candidates, that consist out of too many words. A threshold can be selected based on the length of manually selected keywords. The average over the longest keyphrases (i.e. the keyphrases with the most words) of several documents is a good reference. Taking the maximum instead of the average is not appropriate, since this may prefer some very long keyphrase, which could be categorized as an outlier. Furthermore, past research has shown that the ideal length of a keyphrase due to the human perception is 4.5 words [1], therefore we remove all keyphrases which contains more than 5 words.

Additionally, we can already make use of the graph measures. A very small betweenness of a keyphrase is an indicator for a non-informative keyphrase, since a low value can be interpreted as missing any connection to the concepts of the paper, e.g. senseless keyphrases or unimportant side issues. This notion only holds true due to the graph being small. In bigger graphs (e.g. working on the full text), betweenness may deteriorate to very small numbers anyways. Selecting a threshold isn't as easy and universal as it was for the keyphrase length. Based on some prior experiments, all keyphrase candidates with a normalized betweenness lower than 0.2 were junk and could therefore be safely removed.

3.4 Ranking Keyphrase Candidates

After purging several keyphrase candidates, most of the remaining ones are still not appropriate for describing the underlying concept of the abstract. Therefore it is necessary to obtain a ranking, that pushes the most important concept to the front of the list. Since dealing with abstracts also restricts the number of sensible keyphrases, it is meaningful to only consider the top-n phrases. In order to have two "safe" candidates for top position, we propose to use the keyphrases with the maximal closeness and the maximal betweenness. As described earlier, these should be keyphrases, that describe the underlying methods, technologies or the general topic of the abstract. There are cases, where betweenness and closeness are maximal for the same keyphrase, indicating a very good candidate for a sensible description.

Since one or two keyphrases are usually not enough to sufficiently describe the content of a paper (or an abstract of it), it is feasible to also have a general purpose ranking as well. Sorting the keyphrases by either closeness or betweenness is not useful, since these values quickly deteriorate in the lower ranks (especially with a sparse and small graph). That is why after preselecting the keyphrases with the highest closeness and betweenness, we sort the remaining candidates not only by their closeness values, but also by the average of the edge weights of the path, as described above. As previously explained, closeness is already a strong indicator for good keyphrase candidates. Adding the average edge weight and therefore the notion of connectedness of phrases to the ranking more likely selects those, that form a semantical unit and therefore provide an actual description of concepts or notions of those.

In total, this leads to a ranking of keyphrases, that likely puts the most important conceptual keyphrases on top of the candidate list. In the best case, the 3 to 5 best keyphrases represent the general idea or goal of the paper, its underlying concept or technology and supporting notions for either.

4   Qualitative Evaluation

Due to its nature, our approach does not directly correspond with manually chosen keywords usually supplied with papers. This is due to the fact, that those manually assigned keywords tend to be more general and broader in concept, in order to have a better understanding on where to place that paper thematically. Our method however is focused on retrieving the specific concepts, therefore extracting very specific keyphrases most of the time. Hence, a direct comparison of manually assigned keywords and extracted keyphrases is not sensible.

There are several known benchmark datasets available for keyphrase extraction. These set are usually designed for precision/recall evaluations based on author- or reader-assigned keywords, for comparing the manual work with the automatic result. For our method these datasets are not feasible, since we limited ourselves to work on only the abstract of the documents, therefore having less material available to work with. Furthermore, since our *conceptual keyphrases* differ substantially from the conventional keyphrases extracted by other means, the usually F-Score based evaluation approach is not feasible here. Consequently, we will show the top 5 keyphrases created by our method applied to a number of abstracts and compare it with the top 5 keyphrases generated by Rake and the keyphrases which have been manually selected by the publisher.

Results

Table 1 shows the result for the abstract of the article "Detecting signals of new technological opportunities using semantic patent analysis and outlier detection" [19]. As one can see, the manually selected keyphrase comprise the general

Table 1. Three examples for the results of our approach

Detecting signals of new technological opportunities using semantic patent analysis and outlier detection [19]

Manually selected keywords:
Technological opportunity, Outlier detection, Patent mining, Subject–action–object (SAO) structure, Semantic patent similarity

Rake:
based semantic patent analysis, competitive business environment, syntactically ordered sentences, natural language processing, encode key findings

Our results:
encode key findings of inventions, technological components in a patent, early identification of technological opportunities, technological opportunities, detect new technological opportunities

---

Featherweight Java: A Minimal Core Calculus for Java and GJ [9]

Manually selected keywords:
Compilation, generic classes, Java, language design, language semantics

Rake:
remaining pleasingly compact, extended system formalizes, introduced lightweight versions, enable rigorous arguments, featherweight java bears

Our results:
studies the consequences of extensions, type safety for featherweight java, featherweight java, extend featherweight java, featherweight java bears

---

Learning Users' Interests by Unobtrusively Observing Their Normal Behavior [7]

Manually selected keywords:
Intelligent web agents, learning user preferences, learning by observation, adaptive information retrieval

Rake:
obtaining labeled training instances, intelligent interfaces attempting, present empirical results, easily measured aspects, learn "surrogate" tasks

Our results:
desired output is easily measured, learn a users interests, obtaining labeled training instances, directly labeled each training instances, unobtrusively observing the users behavior

---

topic of *Patent mining, Semantic patent similarity*, and *Outlier detection*. The keyphrases generated by Rake indicate that the document is about *natural language processing* and patent analysis in general. However, a reader would still not get the general idea and motivation of the article.

Looking at our results, we can easily see that the general goal is to *encode key findings of inventions*, based on the *technological components in a patent* and the broader topic is the *early identification of technological opportunities*.

Nevertheless, it is clear that our approach could probably benefit from a better handling of overlapping keyphrases, since mentioning *technological opportunities* in 3 of 5 conceptual keyphrases is not necessary.

The second example is more problematic. While the manually selected keywords give the reader an idea that the article is about Java and language design, the rake keywords leave a lot to be desired and don't make much sense. Our results indicate that the document deals with *featherweight java*, it's *type safety* and the *consequences of extensions*. The keyphrase *featherweight java bears* also appears in the result list of Rake and does not provide much information.

When we look at our results of the third example, we see that our method extracted the keyphrases *unobtrusively observing the users behaviour* and *learn a users interest*, which are the most suitable for describing the document. However, these phrases also appear in the title of the paper, which we did not include in the algorithm's input. Therefore, when using this method for enriching the display of search results, where the title of the document is usually displayed, one could argue that these keyphrases should be omitted since they do not carry any additional information.

5 Conclusion and Outlook

In this paper we presented a graph-based approach for extracting meaningful, conceptual keyphrases from document abstracts with the goal to allow a user to easily apprehend the document's topics. We showed that by using centrality and other graph-related measures, it is possible to generate conceptual keyphrases only from the abstracts of documents, without the need for manually labeled training data, domain knowledge or some kind of corpora. Therefore this technique can be used i.e. to enrich results in a document search engine, so that a user can get a general idea of each result's topics. For further evaluation the quality of our method, we plan to use it to create conceptual keyphrases from abstracts of a number of technical documents and to ask domain experts to judge the quality of the generated keyphrases. In this way we should be able to observe if our keyphrases could be used to describe the thematical concepts of documents. To improve our method more, we plan to extend it by considering overlapping keyphrases as well as identifying syntactical duplicates. Using our graph, we should be able to select i.e. only representative keyphrases which could improve the coverage of the document's topics further.

Acknowledgements. This research is supported by BMWi grants KF3358702KM4, KF2885203KM4.

References

1. Baddeley, A.D., Thomson, N., Buchanan, M.: Word length and the structure of short-term memory. J. Verbal Learn. Verbal Behav. **14**(6), 575–589 (1975)
2. Barla, M., Bieliková, M.: On deriving tagsonomies: keyword relations coming from crowd. In: Nguyen, N.T., Kowalczyk, R., Chen, S.-M. (eds.) ICCCI 2009. LNCS (LNAI), vol. 5796, pp. 309–320. Springer, Heidelberg (2009). doi:10.1007/978-3-642-04441-0_27

3. Beliga, S., Meštrović, A., Martinčić-Ipšić, S.: An overview of graph-based keyword extraction methods and approaches. J. Inf. Organ. Sci. **39**(1), 1–20 (2015)
4. Brandes, U.: A faster algorithm for betweenness centrality*. J. Math. Sociol. **25**(2), 163–177 (2001)
5. Buyukkokten, O., Garcia-Molina, H., Paepcke, A.: Seeing the whole in parts: text summarization for web browsing on handheld devices. In: Proceedings of the 10th International Conference on World Wide Web, pp. 652–662. ACM (2001)
6. Freeman, L.C.: Centrality in social networks - conceptual clarification. Soc. Netw. **1**, 215–239 (1978)
7. Goecks, J., Shavlik, J.: Learning users' interests by unobtrusively observing their normal behavior. In: Proceedings of the 5th International Conference on Intelligent user interfaces, pp. 129–132. ACM (2000)
8. Grothe, L., Luca, E.W.D., Nürnberger, A.: A comparative study on language identification methods. In: Proceedings of the Sixth International Language Resources and Evaluation (LREC 2008), pp. 980–985 (2008)
9. Igarashi, A., Pierce, B.C., Wadler, P.: Featherweight Java: a minimal core calculus for Java and GJ. ACM Trans. Program. Lang. Syst. (1999)
10. Lahiri, S., Choudhury, S.R., Caragea, C.: Keyword and keyphrase extraction using centrality measures on collocation networks. arXiv:1401.6571 (2014)
11. Lossio-Ventura, J.A., Jonquet, C., Roche, M., Teisseire, M.: Yet another ranking function for automatic multiword term extraction. In: Przepiórkowski, A., Ogrodniczuk, M. (eds.) NLP 2014. LNCS (LNAI), vol. 8686, pp. 52–64. Springer, Heidelberg (2014). doi:10.1007/978-3-319-10888-9_6
12. Popova, S., Khodyrev, I.: Ranking in keyphrase extraction problem: is it suitable to use statistics of words occurrences. Proc. Inst. Syst. Program. **26**(4), 123–136 (2014)
13. Popova, S., Kovriguina, L., Mouromtsev, D., Khodyrev, I.: Stop-words in keyphrase extraction problem. In: 2013 14th Conference of Open Innovations Association (FRUCT), pp. 113–121. IEEE (2013)
14. Rose, S., Engel, D., Cramer, N., Cowley, W.: Automatic keyword extraction from individual documents. In: Berry, M.W., Kogan, J. (eds.) Text Mining, pp. 1–20. Wiley, New York (2010)
15. Sarıyüce, A.E., Kaya, K., Saule, E., Catalyürek, U.V.: Incremental algorithms for closeness centrality. In: IEEE International Conference on BigData (2013)
16. Šišović, S., Martinčić-Ipšić, S., Meštrović, A.: Toward network-based keyword extraction from multitopic web documents. In: International Conference on Information Technologies and Information Society (ITIS 2014) (2014)
17. Wang, R., Liu, W., McDonald, C.: Using word embeddings to enhance keyword identification for scientific publications. In: Sharaf, M.A., Cheema, M.A., Qi, J. (eds.) ADC 2015. LNCS, vol. 9093, pp. 257–268. Springer, Heidelberg (2015). doi:10.1007/978-3-319-19548-3_21
18. Xie, Z.: Centrality measures in text mining: prediction of noun phrases that appear in abstracts. In: Proceedings of the ACL Student Research Workshop, pp. 103–108. Association for Computational Linguistics, Stroudsburg (2005)
19. Yoon, J., Kim, K.: Detecting signals of new technological opportunities using semantic patent analysis and outlier detection. Scientometrics **90**(2), 445–461 (2011)

Back to the Sketch-Board: Integrating Keyword Search, Semantics, and Information Retrieval

Joel Azzopardi[1(✉)], Fabio Benedetti[2], Francesco Guerra[2], and Mihai Lupu[3]

[1] University of Malta, Msida, Malta
joel.azzopardi@um.edu.mt
[2] Universitá di Modena e Reggio Emilia, Modena, Italy
{fabio.benedetti,francesco.guerra}@unimore.it
[3] TU Wien, Vienna, Austria
mihai.lupu@tuwien.ac.at

Abstract. We reproduce recent research results combining semantic and information retrieval methods. Additionally, we expand the existing state of the art by combining the semantic representations with IR methods from the probabilistic relevance framework. We demonstrate a significant increase in performance, as measured by standard evaluation metrics.

1 Introduction

By dealing with the satisfaction of the users' information needs, Information Retrieval (IR) is one of the most enduring, challenging and studied topics in the Computer Science research community. In its most basic structure, an IR approach is composed of (a) a meaningful technique for representing data collections and users' queries; (b) a similarity measure for computing the closeness of data and query representations; and (c) an algorithm able to efficiently rank the data collection according to a user's query.

In this paper, we focus on an ad-hoc retrieval task, where the input query is expressed in the form of keywords and the data is a collection of documents. In this scenario, the task of satisfying the users' information needs mainly requires to deal with the ambiguity of the keywords in a query (e.g., terms can be polysemous and refer to different things in different contexts) and to find a measure for effectively ranking documents (usually composed of a large number of terms) according to users' queries (usually composed of a number of terms which is some order of magnitude smaller than the document).

A large number of techniques have been developed for addressing this task, as reported in the related work Section (techniques for representing data, for measuring the similarity of queries and documents and for ranking the results). In this study, we first reproduce a recent experimental study [3] and then extend it by using a recently proposed term weighting method [9]. We explore if the results provided by a classical IR bag-of-word based approach can be improved

© Springer International Publishing AG 2017
A. Calì et al. (Eds.): IKC 2016, LNCS 10151, pp. 49–61, 2017.
DOI: 10.1007/978-3-319-53640-8_5

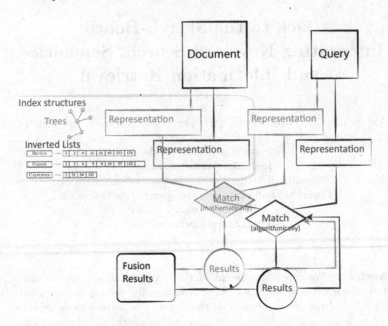

Fig. 1. A typical system architecture combining text and semantic keywords

through the use of some semantic techniques applied on the queries and the document collection.

The architecture of a typical text-semantic system is shown in Fig. 1. The idea is to introduce new semantic representations of the documents and queries. Each representation will provide a ranked list of documents for each input query, which can be considered as a "partial answer" to the query, reflecting the semantics of the representation adopted. A unified answer is then generated by merging all the partial answers. A weighting mechanism can be introduced, as usually done in these approaches, for providing more importance to specific semantic representations.

The starting point of our work is [3], where an annotated query and dataset has been analyzed by taking into account four semantic layers and the weights of document and query vectors have been assigned according to the usual product of Term Frequency (tf) and Inverse Document Frequency (idf). We explore improvements to this approach in two directions: (1) experimenting with new semantic representations of the data; (2) experimenting with different measures for computing the closeness of documents and queries. Concerning the first direction, we started from a subset of the layers analyzed in KE4IR [3], by taking into account only classes and entities referenced in the data. The hypothesis here is that these concepts should reduce the noise generated by spurious information. Additionally, this set of annotations has been improved and extended in two ways. First, as in KE4IR, by relying on PIKES[1] [4] for finding related classes

---

[1] http://pikes.fbk.eu/.

and entities. We refer to this as the *enriched* set of entities and classes. Second, we experiment a refinement and extension of the annotations by exploiting the information about the entities and classes provided by DBpedia. In particular, for most of the terms, DBpedia, through the *abstract* attribute, provides a short textual description of it. The application of AlchemyAPI[2] to the abstracts of the classes allows us to retrieves the entities referenced in them to be used for extending the representation. Concerning the second direction, we experiment the classical BM25 measure [14] and its recently introduced variant [9]. The results obtained from our experiments demonstrate that semantics and specific similarity measures can improve the quality of the results obtained.

Summarizing, the contribution of this paper is:

- we reproduce the work in KE4IR [3];
- we extend the work by introducing new semantic representations of data and queries;
- we change the scoring function from the tfidf to the BM25 [14] and BM25 variant [9].

The rest of the paper is structured as follows. Section 2 introduces some background and related work; Sect. 3 describes the variation of BM25 scoring function adopted in the paper and Sect. 4 shows the results of our experiments. Finally, Sect. 5 sketches out some conclusion and future work.

2   Related Work and Background

Several approaches have been proposed in the literature for providing "semantic annotations" to describe unstructured documents. Depending on the specific goal motivating the approach, the related research community, the existence of a reference vocabulary/ontology of classes and entities, and the output provided and many other criteria, this task has been called in different ways as named entity recognition (NER), semantic annotation, knowledge and information extraction, or ontology population. Some interesting surveys on the topic are provided by Nadeau and Sekine [12], where some approaches for "named entity recognition" are classified on the basis of their learning method and set of features adopted; and Gangemi [5], where 14 tools for Information Extraction are analyzed and classified.

2.1   Probabilistic Relevance Framework

One of the fundamental ways of thinking about information retrieval is the Probabilistic Relevance Framework [14]. It is the so-called *classical* probabilistic model because it is that model that has its roots in the early work of Maron and Kuhns [11] in the early '60s, and later in that of Van Rijsbergen [17] and Spark Jones [7]. Its most conspicuous advocate is however Robertson [15].

---

[2] http://www.alchemyapi.com/.

As pointed out in Fig. 1, in information retrieval (the black boxes), documents and queries are transformed to some common representation and then compared. This comparison, while it may, mathematically, be very precise (e.g. comparing two vectors is well defined and for any distance function we will have a deterministic output) is in reality unavoidably subjected to the uncertainty of language. Mathematically, the only way we can quantify and work with uncertainty are probabilities.

The Probabilistic Relevance Framework (PRF) ranks the documents by the estimated probability that a hidden random variable R takes one of two values (some authors use $1/0$, others $r/\bar{r}$ or even $l/\bar{l}$ to denote *relevance* and *not relevance*). Estimating this probability for information retrieval consists of fundamentally two steps:

1. finding measurable statistics that we consider indicative of relevance (e.g. term frequency, collection frequency)
2. combining these statistics to estimate the probability of a documents relevance to the query

The affability of the PRF derives from the probability ranking principle, first publicly formulated by Robertson [15], but credited by him to private communication with W. Cooper of Univ. of California at Berkeley, and first hinted at by Maron and Kuhns:

"If a reference retrieval system's response to each request is a ranking of the documents in the collections in order of decreasing probability of usefulness to the user who submitted the request, where the probabilities are estimated as accurately as possible on the basis of whatever data has been made available to the system for this purpose, then overall effectiveness of the system to its users will be the best that is obtained on the basis of this data."

The methods developed as a consequence of this principle, while often restricted to statistics that come out of the text, are not bound to this limitation. As the principle states, we can base this calculation on *"whatever data has been made available to the system"*. In the Web domain this freedom has been used to combine, for instance, Roberston's BM25 Relevance Status Value with PageRank [14], thereby defining relevance as a combination of topical relevance and importance in a network.

2.2   Term Weighting

With respect to term weighting schemes, the last 60 years have essentially seen three methods: the initial heuristics of the vector space model, the various probabilistic models (including here Probabilistic Relevance Framework mentioned above but also Language Modelling [13] and Divergence from Randomness [1]), and machine learning approaches [10].

Among them, the Probabilistic Relevance Framework (PRF) and Language Modelling have received most of the attention of the community. While they are conceptually different (most notably with respect to the probability spaces in which they operate), there is a strong relationship between them [8]. This is partially why in this study we only focus on the best known instantiation of the PRF, BM25, and the recently introduced variant BM25$_{VA}$. The other reason is that language modelling methods are very sensitive to parameter tuning, and in our case the relatively small size of the test collection might make the experiments unreproducible to other environments. As Zhai [20] pointed out:

> "This may be the reason why language models have not yet been able to outperform well-tuned full-fledged traditional methods consistently and convincingly in TREC evaluation."

The Probabilistic Relevance Framework, as extensively described most recently by Robertson and Zaragoza [14], assumes that the term frequency of individual words is generated by the composition of two Poisson distributions: one for the occurrence of the term and one of the term being *elite* or not (where by *elite*, Roberston denotes those terms that bear the meaning of documents). However, as the two Poisson distributions are in practice impossible to estimate accurately, the weight of each term is approximated by

$$w_t = log \frac{|D| - df_t + 0.5}{df_t + 0.5} \cdot \frac{tf_t}{k_1 + tf_t}$$

Since BM25 does not use the cosine similarity (there are no vectors), a length normalisation is directly applied on the term frequency component. Thus, a score is computed for each document d and query q as follows:

$$S(q,d) = \sum_{t \in T_d \cap T_q} \frac{(k_3 + 1)tf_q}{k_3 + tf_q} \frac{(k_1 + 1)\overline{tf_d}}{k_1 + \overline{tf_d}} log \frac{|D| + 0.5}{df_t + 0.5} \tag{1}$$

where

$$\overline{tf_d} = \frac{tf_d}{B} \qquad B = (1 - b) + b\frac{L_d}{avgdl}$$

where tf_q and tf_d are the term frequency of a term in the query and the document T_q and T_d are the set of unique terms in the query and the document, $|D|$ is the total number of documents in the collection, L_d is the length of document d (i.e. number of tokens) and $avgdl$ is the average length of a document in the collection.

Recently, Lipani et al. [9], based on the observation that the length normalisation is related to the average term frequency, introduced a variant of B that eliminates the free parameter b:

$$B_{VA} = \frac{avgtf_d}{mavgtf^2} + (1 - mavgtf^{-1})\frac{L_d}{avgdl} \tag{2}$$

where

$$avgtf_d = \frac{1}{|T_d|} \sum_{t \in T_d} tf_d(t) = \frac{L_d}{|T_d|} \quad \text{and} \quad mavgtf = \frac{1}{|D|} \sum_{d \in D} avgtf_d$$

are the average entity frequency and the mean average entity frequency, respectively.

2.3 Concepts

In this paper we exploit classes and entities identified and extracted from the documents to improve the information retrieval process. As in KE4IR [3], PIKES [4] has been adopted as a tool for extracting classes and entities, but our proposal is independent of the tool used. We chose to use the PIKES dataset since it has been constructed with the explicit purpose of experimentation on small collections that present the challenge of the vocabulary gap. This is not the first time that semantic techniques have been used for supporting the IR process. In particular, semantics can support the search task in all the phases of the process: by providing possible interpretations and meanings to documents and queries, and by improving the matching and ranking of queries and documents. In this paper, we extend and improve the document and queries representation, by taking into account classes and entities extracted from them. A similar approach has been followed in other interesting proposals where, for example, a number of research groups worked on extending the Vector Space Model (VSM) by embedding additional type of information in it. The results obtained are not always convincing, in particular in the definition of a theoretically sound interpretation of the resulting vector space. Among the most interesting approaches, there are Tsatsaronis and Panagiotopoulou [16], where a technique exploiting the WordNet's semantic information has been experimented against three TREC collections; and Waitelonis et al. [18], where a generalized vector space model with taxonomic relationships has been introduced. The last approach was experimented, and obtained good performance, against Linked Data collections.

Other approaches exploited semantic techniques for analyzing the dataset according to different perspectives: Gonzalo et al. [6] proposed to use a VSM where the index terms adopted are the synsets extracted from WordNet. Corcoglioniti et al. [3] proposed KE4IR, a system based on the combination of the knowledge provided by four semantic layers in conjunction with the usual bag of word representation of the data.

Finally, semantic techniques have been frequently adopted in keyword search approaches over structured data sources. Yu et al. [19] surveys the main approaches in the area, and [2] exploits semantic techniques for integrating, querying and analyzing heterogeneous data sources.

3 Terms and Concepts

We now discuss how to combine the information provided by the terms and the concepts of the document in answering a query. We say "concepts" to refer uniformly to entities, classes, or a combination of the two. The probabilistic relevance framework does not restrict us to using terms, and therefore we can consider the use of concepts in the scoring value.

Let us therefore consider the representation of a document d as a combination of its terms and concepts:

$$d = (\overbrace{tf(t_1), tf(t_2), ..., tf(t_{|T|})}^{\text{term frequencies}}, \overbrace{ef(e_1), ef(e_2), ..., ef(e_{|E|})}^{\text{concept frequencies}})$$

where T is the set of unique terms in the collection and E is the set of unique concepts in the collection and $tf(t_i)$ and $ef(e_j)$ denote the (term) frequency of a term t_i and (concept) frequency of a concept e_j.

Even though d looks like a typical frequency vector, directly applying the reasoning behind BM25 to the new vector does not make sense because the terms and concepts do not share the same probability space. However, building on the BM25 weighting schemes in [9,14], we can define a specific score, which makes the same assumptions as the traditional BM25 scoring method above:

$$S_E(q,d) = \sum_{e \in E_d \cap E_q} \frac{(k_3+1)ef_q}{k_3 + ef_q} \frac{(k_1+1)\overline{ef_d}}{k_1 + \overline{ef_d}} \log \frac{|D| + 0.5}{df_e + 0.5} \tag{3}$$

where

$$\overline{ef_d} = \frac{ef_d}{B_e} \qquad B_e = (1-b) + b\frac{L_d^e}{avgdl_e}$$

where all the elements are the direct translations of the statistics based on terms to statistics based on concepts.

$$B_{VA}^e = \frac{avgef_d}{mavgef^2} + (1 - mavgef^{-1})\frac{L_d^e}{avgdl_e} \tag{4}$$

where

$$avgef_d = \frac{1}{|E_d|} \sum_{e \in E_d} ef_d(e) = \frac{L_d^e}{|E_d|} \quad \text{and} \quad mavgef = \frac{1}{|D|} \sum_{d \in D} avgef_d$$

are the average concept frequency and the mean average concept frequency.

This new score component $S_E(q,d)$ can be linearly combined with the term-based score component $S(q,d)$ from Sect. 2:

$$\mathbf{S}(q,d) = S(q,d) + \lambda S_E(q,d) \tag{5}$$

To note that we do not use here the more common linear combination (λ)—$(1 - \lambda)$ for two reasons: first, the addition above is calculated based on the extended vector representation of the document, as noted at the beginning of this section. λ is introduced as a parameter to give more or less importance to this additional information. Second, it seems counter intuitive to force the text and concept components to play against each other (i.e. by increasing λ on one of them to automatically decrease $1 - \lambda$ for the other). There is no theoretical nor practical consideration for which we would do that here, since we do not need our score to be constrained within a given interval.

4   Experimental Results

We performed 4 sets of experiments, namely:

1. Using terms alone comparing traditional BM25 (standard B) with the variation B_{VA} introduced in [9], as well as the baseline in [3];
2. Using terms (as in 1 above) *after* applying filtering based on concepts;
3. Combining ranking of terms and concepts as defined in Eq. 5; and
4. Combining ranking of terms and concepts as in 3 *after* applying filtering based on concepts.

The evaluation has been performed against the dataset developed in [18]. The dataset consists of 331 articles from the yovisto blog, each one composed of 570 words in average. The articles are annotated (83 annotations per article in average). The queries have been inspired by the search log and have been manually annotated.

The metrics used are the same as those used in Corcoglioniti et al. [3], that we assume as baseline, namely: Precision at three different positions—namely *P@1*, *P@5*, *P@10*, Normalised Discounted Cumulative Gain (NDCG) and Mean Average Precision (MAP). The latter two are computed after the first ten retrieved documents, and for the entire ranked list. In all subsequent plots, a solid horizontal line indicates the best performing KE4IR run as reported in [3], while a dotted line indicates the best text-only run as reported in the same paper.

4.1   Retrieval Using Terms Alone

In this set of experiments, we did not utilise any semantic information whatsoever. We performed retrieval using BM25 where we compared the use of the standard B as defined by Eq. 1 with B_{VA} as described in Eq. 4. We used the following parameters: $k1 = 1.2$, $k3 = 0$ and $b = 0.75$ (b is only used in the standard B). The values for $k1$ and b are commonly used values according to [9]. We set $k3$ to 0 to remove possible effects from variations in keyword frequencies within the query itself.

Figure 2 shows the results. One can note that on the whole, the use of standard B produces slightly better results than using B_{VA}. As expected, the use of BM25 on text-only representations does not outperform the best KE4IR configuration described in [3] that uses semantic information. However, BM25 outperforms the text-only KE4IR system on all metrics except for $P@1$ and $P@5$.

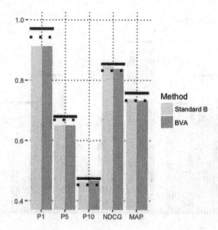

Fig. 2. Results from text data alone, according to Eq. 3 (BVA) and traditional BM25 (Standard B)

4.2 Retrieval Using Terms and Filter on Concepts

The configuration used here is very similar to that described in Sect. 4.1. However, prior to comparing the documents with the query, the document list was filtered to remove those documents that do not contain the concepts present in the query. We compare the use of different sets of concepts, namely: (a) classes, (b) entities, (c) classes and entities, (d) classes extracted through the application of the service provided by AlchemyAPI from the abstract properties of the corresponding DBpedia entries, (e) entities extracted with AlchemyAPI from the abstract properties of the corresponding DBpedia entries, (f) classes and entities extracted from the abstract properties of the corresponding DBpedia entries, (g) classes extracted from the data enriched by using PIKES, (h) entities extracted from the data enriched by using PIKES and (i) classes and entities extracted from the data enriched by using PIKES.

As shown in the results (Fig. 3), most runs outperform the best KE4IR configuration on $P@5$ and $P@10$. The best $P@5$ score is obtained when using B_{VA} and filtering on concepts. On the other hand, the best $P@10$ is obtained when using standard B, and applying filtering using concepts from enriched classes and entities. Despite obtaining relatively high $P@5$ and $P@10$, this system scores lower than KE4IR on $P@1$, $NDCG$ and MAP. This indicates that despite tending to retrieve more relevant documents in the first 5 and 10 positions, its ranking tends to be worse than KE4IR.

4.3 Retrieval Using Combined Ranking of Terms and Concepts

This experiment uses the document-query similarity function defined in Eq. 5, comparing the effect of standard B, and B_{VA} for with various values for λ— ranging from 0.2 to 1.4 in intervals of 0.2. Different runs are performed for the different sets of concepts listed in Sect. 4.2.

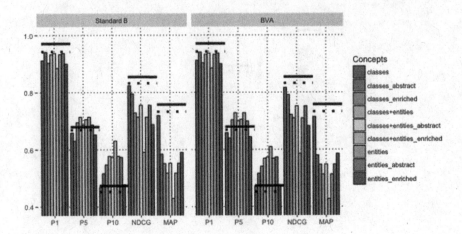

Fig. 3. Results ranking by text data alone, according to Eq. 3 (BVA) and traditional BM25 (Standard B), but filtering on concepts

Results (Fig. 4) indicate that this configuration performs worse that the previous set-up (textual retrieval with semantic filtering) as far as $P@n$ is concerned. On the other hand, there are cases (using entities concepts, standard B, and $\lambda = 1.4$) where $P@10$, and MAP scores exceed the corresponding KE4IR scores. This may imply an improved ranking.

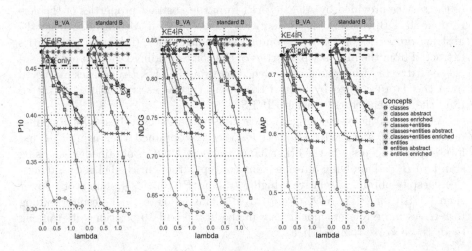

Fig. 4. Results combining text and concepts according to Eq. 3 (BVA) and traditional BM25 (Standard B)

4.4   Retrieval Using Combined Ranking of Terms and Concepts, and Filter on Concepts

This configuration is similar to the setup described in Sect. 4.3, but with the addition of concept filtering (as described in Sect. 4.2). The results obtained here are overall better than that obtained when using textual and conceptual ranking without filtering. Configurations involving B_{VA} and entities keywords manage to produce a $P@5$ score (0.7375) that exceeds the $P@5$ in KE4IR and all other runs. There are also various configurations that exceed the KE4IR $P@10$ score. However, the $P@1$, $NDCG$ and MAP scores are generally lower than the ones obtained without concept filtering.

4.5   Discussion

The experiments allow us to draw the following conclusions:

- The best results were obtained on P@5 and P@10, significantly improving the current state of the art on the provided test collection.
- By considering the top-heavy metrics (P@1 and MAP), the experiments show that it is extremely difficult to significantly improve on the existing results.
- The increased performance in precision obtained by our technique does not correspond to an increase in the $NDCG$ and MAP scores, thus meaning that a larger number of correct documents is associated to a worst ranking of them.
- The main benefit from the adoption of concepts is then related to the filtering of the documents. The results show that in most cases they introduce more noise than utility into the ranking ($NDCG$ and MAP decrease with the increase of λ in Figs. 4 and 5). The small increase in performance for some

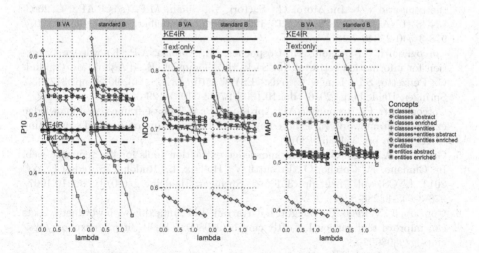

Fig. 5. Results combining text and concepts according to Eq. 3 (BVA) and traditional BM25 (Standard B) as well as filtering on concepts

configurations refers to settings where no filtering has been performed, thus
confirming our findings.
- Due to the small dataset and number of queries evaluated, the result cannot
be generalized out of this domain.
- In this particular domain, the variation of BM25 introduced does not improve
the scores.

5   Conclusion

We have extensively explored the use of concepts in an ad-hoc retrieval task, by
combining them with term-based relevance in three ways: as a filter, as a ranking
contributor, and as both a filter and a ranking contributor. We have explored 9
different interpretations of 'concepts' and concluded that, with reference to the
dataset studied, the main benefit we achieve by taking into account concepts
concerns the reduction of the vector space, where the irrelevant documents are
filtered out. On the other hand a small benefit (and in some case a decrease in
the performance) is measured if we consider the rank. Nevertheless, the problem
needs more evaluation, with other, larger datasets, to draw general conclusions.

Acknowledgments. This research is partially supported by the ADmIRE Project
(FWF P25905-N23) project and the COST IC1302 KEYSTONE Action.

References

1. Amati, G., Van Rijsbergen, C.J.: Probabilistic models of information retrieval
based on measuring the divergence from randomness. TOIS (2002)
2. Bergamaschi, S., Guerra, F., Vincini, M.: A peer-to-peer information system for
the semantic web. In: Moro, G., Sartori, C., Singh, M.P. (eds.) AP2PC 2003.
LNCS (LNAI), vol. 2872, pp. 113–122. Springer, Heidelberg (2004). doi:10.1007/
978-3-540-25840-7_12
3. Corcoglioniti, F., Dragoni, M., Rospocher, M., Aprosio, A.P.: Knowledge extraction for information retrieval. In: Sack, H., Blomqvist, E., d'Aquin, M., Ghidini,
C., Ponzetto, S.P., Lange, C. (eds.) ESWC 2016. LNCS, vol. 9678, pp. 317–333.
Springer, Heidelberg (2016). doi:10.1007/978-3-319-34129-3_20
4. Corcoglioniti, F., Rospocher, M., Aprosio, A.P.: A 2-phase frame-based knowledge
extraction framework. In: Proceeding of ACM Symposium on Applied Computing
(SAC 2016), pp. 354–361 (2016)
5. Gangemi, A.: A comparison of knowledge extraction tools for the semantic web.
In: Cimiano, P., Corcho, O., Presutti, V., Hollink, L., Rudolph, S. (eds.) ESWC
2013. LNCS, vol. 7882, pp. 351–366. Springer, Heidelberg (2013). doi:10.1007/
978-3-642-38288-8_24
6. Gonzalo, J., Verdejo, F., Chugur, I., Cigarrán, J.M.: Indexing with wordnet synsets
can improve text retrieval. CoRR cmp-lg/9808002 (1998). http://arxiv.org/abs/
cmp-lg/9808002
7. Jones, K.S.: Information Retrieval Experiment. Butterworths (1981)
8. Lafferty, J.D., Zhai, C.: Probabilistic relevance models based on document and
query generation. In: Language modeling and information retrieval (2003)

9. Lipani, A., Lupu, M., Hanbury, A., Aizawa, A.: Verboseness fission for BM25 document length normalization. In: Proceeding of ICTIR (2015)
10. Liu, T.Y.: Learning to rank for information retrieval. Found. Trends Inf. Retrieval **3**(3), 225–331 (2009)
11. Maron, M.E., Kuhns, J.L.: On relevance, probabilistic indexing and information retrieval. J. ACM **7**(3), 216–244 (1960)
12. Nadeau, D., Sekine, S.: A survey of named entity recognition and classification. Lingvisticae Investigationes **30**(1), 3–26 (2007)
13. Ponte, J.M., Croft, W.B.: A language modeling approach to information retrieval. In: Proceedings of the 21st Annual International ACM SIGIR Conference on Research and Development in Information Retrieval, SIGIR 1998, pp. 275–281, NY, USA (1998). http://doi.acm.org/10.1145/290941.291008
14. Robertson, S., Zaragoza, H.: The probabilistic relevance framework: BM25 and beyond. Found. Trends Inf. Retrieval **3**(4), 333–389 (2009)
15. Robertson, S.E.: The Probability Ranking Principle in IR. Journal of Documentation 33(4) (1977)
16. Tsatsaronis, G., Panagiotopoulou, V.: A generalized vector space model for text retrieval based on semantic relatedness. In: Lascarides, A., Gardent, C., Nivre, J. (eds.) EACL 2009, 12th Conference of the European Chapter of the Association for Computational Linguistics, Proceedings of the Conference, Athens, Greece, March 30 - April 3, 2009, pp. 70–78. The Association for Computer Linguistics (2009). http://www.aclweb.org/anthology/E09-3009
17. Van Rijsbergen, C.J.: Information Retrieval, 2nd edn. Butterworth, London (1979). http://www.dcs.gla.ac.uk/Keith/Preface.html
18. Waitelonis, J., Exeler, C., Sack, H.: Linked data enabled generalized vector space model to improve document retrieval. In: Proceeding of 3rd International Workshop on NLP & DBpedia 2015, co-located with ISWC (2015)
19. Yu, J.X., Qin, L., Chang, L.: Keyword Search in Databases. Morgan & Claypool Pub, Synthesis Lectures on Data Management (2010)
20. Zhai, C.: Statistical language models for information retrieval a critical review. Found. Trends Inf. Retr. **2**(3), 137–213 (2008). http://dx.doi.org/10.1561/1500000008

Topic Detection in Multichannel
Italian Newspapers

Laura Po[1]([⊠]), Federica Rollo[1], and Raquel Trillo Lado[2]

[1] Dipartimento di Ingegneria "Enzo Ferrari",
Università di Modena e Reggio Emilia, Modena, Italy
laura.po@unimore.it, federica.rollo@libero.it
[2] Departamento de Informática e Ingeniería de Sistemas,
Universidad de Zaragoza, Zaragoza, Spain
raqueltl@unizar.es

Abstract. Nowadays, any person, company or public institution uses
and exploits different channels to share private or public information
with other people (friends, customers, relatives, etc.) or institutions. This
context has changed the journalism, thus, the major newspapers report
news not just on its own web site, but also on several social media such as
Twitter or YouTube. The use of multiple communication media stimu-
lates the need for integration and analysis of the content published glob-
ally and not just at the level of a single medium. An analysis to achieve
a comprehensive overview of the information that reaches the end users
and how they consume the information is needed. This analysis should
identify the main topics in the news flow and reveal the mechanisms of
publication of news on different media (e.g. news timeline). Currently,
most of the work on this area is still focused on a single medium. So, an
analysis across different media (channels) should improve the result of
topic detection. This paper shows the application of a graph analytical
approach, called Keygraph, to a set of very heterogeneous documents
such as the news published on various media. A preliminary evaluation
on the news published in a 5 days period was able to identify the main
topics within the publications of a single newspaper, and also within the
publications of 20 newspapers on several on-line channels.

Keywords: Clustering · Topic detection · News cycle · News tracking ·
Cross-channel publication · Social media

1 Introduction

In the last decade, more and more newspapers have begun using the Internet as a
tool for spreading news. Printed newspapers continue to be used but editors are
increasingly distributing their newspapers' content over several delivery channels
on the Internet due to improved timeliness. In most cases, they are re-purposing

The research presented in this paper was partially funded by Keystone Action COST
IC1302.

A. Calì et al. (Eds.): IKC 2016, LNCS 10151, pp. 62–75, 2017.
DOI: 10.1007/978-3-319-53640-8_6

content from the printed editions in various electronic editions, on their web sites and on social media [11]. Thus, according to ISTAT research[1], the percentage of Italians reading newspapers on-line was 11% of the total of Italian people using the Internet, while in 2014 this percentage rose to 32.2%. France and Poland have similar rates, while Finland and Sweden have more digital readers (80%).

Recently, social networks have gained a very important role in the dissemination of news, because they allow a greater share of news than websites and are more timely to provide updates. In fact, it is very common that a newspaper publishes several updated versions of the same piece of news on the same day. Therefore, as well as printed newspapers and their websites, most journals also use social networks, specially Facebook and Twitter. So, it is interesting to delineate how and how much Web and social networks are used to disseminate news content.

The goal of this paper is to analyze published news to determine whether there exist correlations among news published by different newspapers on different channels. In order to do that, we have adapted the Keygraph algorithm [9] for topic detection on multiple communications media. The idea was to devise a new approach that can examine all the contents coming from different media (currently: web sites, Facebook, Twitter) instead of considering an analysis focused on a single media. Besides, we have performed a preliminary evaluation on a 5 days period of the news collected on December 2015, that were published on-line by the 20 most popular Italian newspapers.

Clustering news published on different communication media is difficult since different styles are used in different media or channels. Thus, the news reported on the web sites usually contain several phrases, a title (usually short) and a long description, while the posts about the same pieces of news on social media contain few words and other kinds of information such as hashtags and links. On the one hand, we can exploit the implicit information encoded in the hashtags and in the links toward other pieces of news. On the other hand, the preprocessing and the clustering techniques need to be configured and modified based on the input text.

This paper is organized as follows. In the next section, we review related work, Sect. 3 presents the Keygraph topic detection algorithm and the modifications introduced to harmonize the topic detection when applied on heterogeneous sources. Then, in Sect. 4, we test the impact and accuracy of Keygraph on Italian newspaper publications. Finally, we summarize our research and highlight some future directions in Sect. 5.

2   Related Work

Documents describing the same news usually contain a lot of words in common and have similar sets of keywords. Topic/Event Detection and Tracking [5,8] is a research field aiming to develop technologies that search, organize and structure

---

[1] ISTAT http://tinyurl.com/jc5sfc8.

textual materials from a variety of broadcast news media. With the explosion of social networks such as Facebook and Twitter, techniques for event detection were adapted to consider streams of shorter documents (entries) produced with a higher frequency (hours or minutes vs days) [2]. Moreover, new challenges arise: dealing with a higher level of grammatical errors, incorrect spelling, etc. The topics or events extracted from the different collections are usually used to characterize the items of the collection and make recommendations to users. Thus, TMR [6] is an semantic recommender system that takes as input a Topic Map generated by TM-Gen and a profile of a user and outputs a list of items that the user could be interested in. The adapted version of Keygraph [9] described in this paper has the same purpose as TM-Gen, i.e., extracting information from a set of pieces of news and representing them as a Topic Map. Nevertheless, our proposal not only considers the news on the website of the newspaper but it also takes into account the entries published on Twitter and Facebook. Moreover, TM-Gen only considers an information source (the news published by the Spanish newspaper "El Heraldo de Aragón") for performing experiments, while the proposal described in this paper has integrated the information from different media companies (21 Italian newspapers).

3   Keygraph Adapted for the Multichannel Analysis

The aim of this paper is to carry out an analysis on the news published by the main Italian newspapers: clustering the news around the main topics to understand correlations and make comparisons between news published on different channels and different newspapers. The analysis of news has been made using the Keygraph algorithm for automatic indexing of documents. Keygraph explicitly incorporates word co-occurrence in topic modeling and it has been demonstrated to have scalable and good performances, similar to that of topic modeling solutions (such as GAC and LDA-GS), on a large noisy social media dataset [9].

An event is "a specific thing that happens at a specific time and place" [1]. It may be composed of many sub-events, each of them at a finer level of granularity. For example, the event of the Spanish election occurred in December 2015 covers a broad range of topics: the voter turnout, the announcement of the winner, the reaction of the winner and the opposition, the risk of ungovernability, etc. All of these are sub-events related to the Spanish election event. News or events can be described by a set of terms, representing the asserted main point in the document. Documents describing the same event usually contain similar sets of keywords. Therefore, in order to detect the topic of the news and to cluster similar news, it is crucial to extract meaningful keywords and to discharge unessential words from news text.

Keygraph [9] is an algorithm based on the segmentation of a graph, whose goal is to identify events and clusters around events. Keygraph applies a community detection algorithm to group co-occurring keywords into communities. Each community is formed by a constellation of keywords that represents a topic.

The similarity between a community and a document is computed to rank similar documents. The original Keygraph algorithm[2]) was modified to improve the results of indexing, giving consideration to the hashtags and URLs in news text. The algorithm uses a configuration file that is provided as input and contains numerical parameters useful for clustering (the upcoming words written in italics refer to configuration parameters).

In the following, the original Keygraph algorithm phases and the modifications that have been introduced in order to deal with heterogeneous documents (news published on different channels) are described.

3.1   Building the Keygraph

The first phase focuses on extracting keywords from documents, which represent pieces of news, and building a graph considering the co-occurrence of keywords.

The body of a document (content, text describing the piece of news) is the principal component and it is analyzed to extract keywords: each word of the body is stemmed and is considered if and only if it does not appear in a stop-word list (a list of very commonly used words irrelevant for searching purposes). We used the Italian version of the Snowball stemmer[3]. Each keyword k_i is characterized by a base form, that is the root of the word (the result of the stemmer), its term frequency TF (how many times the keyword appears in the document), its document frequency DF (how many times the keyword appears in all documents) and the inverse of its document frequency IDF.

The TF is initialized to *text-weight* value, given as input in the configuration. At this stage, each document is represented by a set of keywords. Documents that have less than *doc_keywords_size_min* keywords are removed.

After that, a node n_i is created for each unique keyword in the dataset. Nodes with low DF or high DF are filtered. An edge $e_{i,j}$ between nodes n_i and n_j is added if k_i and k_j co-occur in the same document. Edges are weighted by how many times the keywords co-occur (DF document frequency of the edge). Edges linking keywords that co-occur below some minimum threshold (*edge_df_min*) are removed. Edges linking keywords that almost always appear together are also removed.

For each remaining edge, conditional probabilities (CP) $p(k_i|k_j)$ and $p(k_j|k_i)$ are computed. For $e_{i,j}$, the conditional probability of the occurrence $p(k_i|k_j)$ is the probability of seeing k_i in a document if k_j exists in the document. The conditional probability is computed in the following way:

$$p(k_i|k_j) = \frac{DF_{i \cap j}}{DF_j} = \frac{DF_{e_{i,j}}}{DF_j}$$

Finally, nodes without edges are removed.

---

[2] The code of the version 2.2 of March 2014 is available on-line at http://keygraph.codeplex.com/.
[3] http://snowball.tartarus.org/.

3.2   Extracting Topic Features

The second stage involves the extraction of communities within the Keygraph created.

The graph appears as a network of interconnected keywords; here some nodes have a stronger connection with others. These groups of interconnected nodes, called connected components, are identified. Each connected component must contain a number of nodes between *cluster_node_size_min* and *cluster_node_size_max*. If the number of nodes is greater than the threshold, the edges with low CP are deleted.

Within these groups, the communities need to be identified. A useful measure for this purpose is the betweenness centrality, an indicator of a node's centrality in a network. The betweenness centrality of an edge is defined as the number of shortest paths for all pairs of nodes of the network that pass through that edge. Of course the edges that connect different communities have a very high betweenness centrality score, since the shortest routes connecting pairs of nodes of different communities will have to pass necessarily by those arcs (edges).

In each connected component it is necessary to identify the edge with the highest betweenness centrality score, through a breadth-first search. This edge is removed from the graph. If two edges have the same score of betweenness centrality, the one with lower DF is removed. Before removing the edge, its value of conditional probabilities is considered. If the CP of the edge is above the threshold *edge_cp_min_to_duplicate*, then the edge and its corresponding nodes are duplicated. In this way, a node might occur in more than one community. This process is repeated until there is no edge with a high betweenness centrality score. After removing all the edges that interconnect different communities, we identify for each community a topic. The topic is characterized by the keywords of the community (the feature vector f_t). So, each community can be seen as a particular document. The documents similar to this "community document" can be clustered together, creating a document cluster.

3.3   Assigning Topics to Documents

The probability that a topic t is associated with a document d is calculated by considering the cosine similarity of d with respect to the feature vector f_t, as follows:

$$p(t|d) = \frac{cosine(d, f_t)}{\sum_{t' \in T} cosine(d, f_{t'})}$$

The weight of each keyword of the feature vector f_t is calculated using the TF-IDF function. This function increases proportionally to the number of times that the word is used in the document, but grows in inverse proportion with the frequency of the term in the collection. So, it gives more importance to the terms that frequently appear in the document, but are quite rare in the collection [7]. The TF-IDF function can be decomposed in two factors: $tf_{i,j}$ and idf_i. $tf_{i,j}$ is the number of occurrences of the term t_i in the document d_j; while idf_i represents the overall importance of the word in the collection and is calculated as follows:

$$idf_i = \frac{1}{ln2} ln\left(\frac{|D|}{DF_i}\right)$$

where $|D|$ is the number of documents in the collection and DF_i is the total number of occurrences of the term t in all documents. The sum of the TF-IDF functions calculated for each node in a community is called *vector size*. The vector size is calculated for the community, for the document and for the set of keywords that are shared between the community and the document. The cosine similarity is the ratio between the latter and the product of the first two. For each document the cosine similarity is computed with respect to each community and its value is compared with the *doc_sim2Keygraph_min* threshold: if the similarity between a document and a topic is greater than this parameter, the topic is assigned to the document. So, similar documents form clusters.

A document may be assigned to multiple topics, unless "hard clustering" (forcing the assignment of a document to the topic with the greatest cosine similarity) is specified. If no "hard clustering" is performed, there may be a significant overlap between the sets of documents in different document clusters. So, a merging operation is performed if the following equation is verified:

$$\frac{intersect}{min(|DC1|, |DC2|)} \geq cluster\_intersect\_min$$

where intersect is the number of common documents and $|DC1|$ and $|DC2|$ are the number of documents that are part of the first and the second document cluster. After this final step, the algorithm created a set of document clusters. Documents within the same cluster are about the same topic.

3.4   Modification to the Algorithm

Initially, minor changes were made to cluster news coming from different channels. These minor changes regard increasing the importance of hashtags, deleting mentions and names of authors from the news text, and defining proper configuration parameters. Hashtags, textual tokens prefixed by hash marks (#), are very useful for our purpose, since they are used as proxies for topics. Each news post can contain several hashtags (especially Twitter posts). In Keygraph, hashtags are extracted, splitted in a list of words, and added to the set of keywords describing the document (if they are not considered stopwords). The hashtag segmentation regards finding the best way of splitting an input string into words. In literature, empirical methods and supervised or unsupervised techniques based on multiple corpora are available for word segmentation. Posts about the same topic may use different, but still similar hashtags. The hashtag segmentation is useful to identify the correlation between these posts. For example, two posts the first containing "#expo2015" and the second "#expoMilano" share the keyword "expo".

In addition to these changes, a major improvement was the implementation of a mechanism that would allow the consideration of link between different documents. Each post can contain one or more URLs, i.e. links to external resources.

Generally, posts on Twitter and Facebook contain two links: one connects to the news on the social network and the other leads to the web site page of the newspaper in which the same news is published. A URL can also be used to link to a previous version of the news regarding the same topic. Moreover, some URLs are links to multimedia contents (such as Youtube videos). In a few cases, URLs are links to "general" pages. This kind of URLs do not connect to a specific news thus we will not consider them. The important point for our purposes is that posts that share a link are strongly correlated, thus we can suppose they are about the same topic. So, we know in advance that these posts should be part of the same document cluster.

All the modifications introduced have led to the implementation of three extended versions of Keygraph, that are variants of the original algorithm [9]:

- Keygraph 1 is a variant of Keygraph that eliminates the authors' names and mentions from the news posts, extracts and analyses the hashtags, uses three different configuration parameters according to each publication channel; and increases the weight of hashtags by doubling their term frequencies.
- Keygraph 2 is a variant of Keygraph 1 that creates arcs and, if necessary, nodes for each keyword of each pair of news with at least one link in common; and doubles the weight of these arcs.
- Keygraph 3 is a variant of Keygraph 2 that, in the final phase of the document clusters creation, forces news that shared a link to appear in the same cluster, i.e., it adds in each cluster news that share at least one link with other news in the cluster.

Newspaper	Paper editions	Digital editions	Total circulation
Corriere della Sera	368 981	95 447	464 428
Repubblica (La)	323 525	58 709	382 234
Sole 24 Ore (Il)	200 155	115 366	315 521
Gazzetta dello Sport (La)	203 516	21 042	224 558
Stampa (La)	214 461	7 198	221 659
Messaggero (Il)	137 678	4 510	142 188
QN – Il Resto del Carlino	122 513	1 234	123 747
Corriere dello Sport - Stadio	121 128	1 272	122 400
Giornale (Il)	103 658	2 115	105 773
Avvenire (L')	103 985	1 578	105 563
QN – La Nazione	98 812	1 094	99 906
Tuttosport	94 970	818	95 788
Libero	75 301	886	76 187
Italia Oggi	54 166	18 157	72 323
Gazzettino (Il)	66 163	4 276	70 439
Fatto Quotidiano (Il)	50 763	13 621	64 384
Secolo XIX (Il)	57 068	1 208	58 276
Tirreno (Il)	56 639	1 539	58 178
Mattino (Il)	50 946	2 429	53 375
QN – Il Giorno	50 597	232	50 829

Fig. 1. The list of the 20 most popular Italian newspapers.

4   Evaluation

The algorithm has been tested on the posts published in different time slots on three channels (website, Facebook and Twitter) by 21 italian newspapers: the 20 most popular Italian newspapers and the Italian multimedia information agency (Agenzia Nazionale Stampa Associata -ANSA[4]-). The list of the newspapers in daily periodicity considered in the experiments and their circulation are shown in Fig. 1[5]. The circulation of a newspaper is the number of copies it distributes on average per day. Circulation could be greater than the number of copies sold, since some newspapers are distributed without cost to the readers. The number of readers is usually higher than the circulation because of the assumption that a copy of a newspaper is read by more than one person.

Considering the set of news published between 20th and 22th December 2015 by all the newspapers on the three channels (11423 news in total), we performed several tests to find the best configuration parameters for Keygraph. For example, decreasing the similarity threshold between a document and the community, more documents are clustered together, but the precision value obtained for the algorithm also decreases. We tested two values of the $doc\_sim2Keygraph\_min$ threshold: 0.18 and 0.30. With a 0.18 threshold, we increase the percentage of documents clustered (from 15% to 27%) but the precision decreases (from 75% to 63%). With a 0.30 value, the percentage of documents decreases (from 14% to 8%) but the precision increases (from 83% to 94%). Changes on other parameters

	Site	Facebook	Twitter
TEXT_WEIGHT	0.6	0.8	1
KEYWORDS_WEIGHT		1	
HASHTAG_WEIGHT		2	
HARD_CLUSTERING		false	
NODE_DF_MIN		2	
NODE_DF_MAX*		0.04	
EDGE_CORRELATION_MIN*		0.03	
EDGE_DF_MIN		3	
DOC_KEYWORDS_SIZE_MIN	3	2	2
DOC_SIM2KEYGRAPH_MIN		0.3	
CLUSTER_NODE_SIZE_MAX		1000	
CLUSTER_NODE_SIZE_MIN		2	
CLUSTER_INTERSECT_MIN		0.65	
TOPIC_MIN_SIZE		2	
EDGE_CP_MIN_TO_DUPLICATE		1	
CLUSTERING_ALG		betweenness	

Fig. 2. Configuration parameters.

---

[4] http://www.ansa.it/.

[5] The average circulations of each newspaper refer to February 2015 as reported by the Italian Federation of Newspaper Publishers (Federazione Italiana Editori Giornali available at http://www.fieg.it).

also affect the values of precision, accuracy and recall. Figure 2 shows the configuration parameters that we considered the best. These parameters were used in the two tests that have been conducted for the three versions of Keygraph shown in Sects. 4.3 and 4.4.

4.1   Performance Measures

The evaluation of Keygraph has been conducted by manually identifying true positives (TP), false positives (FP), false negatives (FN) and true negatives (TN) within every document cluster. After that, the following performance measures were calculated:

$$accuracy = \frac{TP + TN}{TP + TN + FP + FN}; \ precision = \frac{TP}{TP + FP}; \ recall = \frac{TP}{TP + FN};$$

TP and FP are evaluated on the list of documents within the document cluster. While FP and FN has to be judged on documents that have not been selected by the algorithm to be part of the cluster. Since it is not feasible to evaluate all the documents that are not selected within a document cluster, we identified two possibilities to retrieve a reasonable number of documents: (1) to recover documents that shared a reasonable number of keywords with the document cluster[6], and (2) to retrieve documents that shared at least one URL link with one document in the cluster.

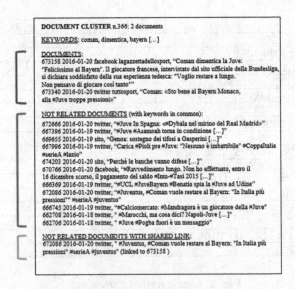

Fig. 3. Keygraph evaluation - an example.

---

[6] As the content available on web sites and Facebook news is greater than the content on Twitter news, the number of shared keywords is different according to the channel: 5 if news is published on a Website or on Facebook, 3 if news is published on Twitter.

A clarifying example is shown in Fig. 3. The blue color indicates the news that are selected by Keygraph to be part of the cluster. The cluster is described by the keywords: "coman", "bayern", "dimentica" (meaning "forgot"), etc. The topic is about the new engagement of the player Coman in the Bayern Monaco team. The two clustered documents are true positives, and no false positives are detected. The red part identifies the news that share a reasonable number of keywords with the set of keywords of the cluster. Here, we can detect that the news with ID number "672086" is related with the topic of the cluster, so this news is a classify as a FN, while all the other news are TN. The green color indicates the news that share at least one URL link with the other documents in the cluster. The detected FN refers to the same news "672086". Note that false negatives are considered two, even if the detected news is the same. In this small example, we got $TP = 2$, $FP = 0$, $FN = 2$, $TN = 10$.

4.2   Evaluation of the Impact

We decide to evaluate the impact of our approach as the number of new correlations created by the algorithm once the documents clusters are built. The correlations can be divided into: correlations among news published on the same channel, correlations among news published on different channels, correlations among news published on the same newspaper, correlations among news published on different newspapers.

The total number of correlations in a document cluster is computed by the formula $Correlations = \binom{N}{2}$ where N is the number of documents in the cluster. For each cluster the number of news published on the same channel is also calculated for each channel (N_{site}, $N_{Facebook}$ and $N_{Twitter}$). If the number of news on a channel is not equal to 0 or 1, the next formulas are used to find the number of correlations between news published on the same channel:

$$Corr_{site} = \binom{N_{site}}{2}; \; Corr_{facebook} = \binom{N_{Facebook}}{2}; \; Corr_{twitter} = \binom{N_{Twitter}}{2}$$

The total number of correlations between the news published on the same channel and on different channels are: $Corr_{sameChannel} = Corr_{site} + Corr_{facebook} + Corr_{twitter}$ and $Corr_{differentChannels} = Correlations - Corr_{sameChannel}$, respectively. Moreover, the same type of calculus is adopted to find correlations among news published by one newspaper or different newspapers.

Finally, the evaluation of the impact is compared with respect to a baseline, called *link cluster*. The *link cluster* is built taking into account only the URL links contained in the news: the news that share at least a URL link are joined in the same document cluster. Thefore, the *link cluster* produces a set of document clusters in which each news share at least a URL link with another news in the same cluster.

4.3   Test 1 - Multichannel Publishing by a Single newspaper

This test has been executed on the news published by La Repubblica on all channels. This test set contains 2430 news (125 published on the Website -5%-,

1307 on Facebook -54%- and 998 on Twitter -41%-) and 1404 links connecting the news in the test set. Besides, 112 documents share at least one link with other document.

The same configuration file was used to run the three versions of the algorithm and the results are shown in Table 1. In the first phase of the algorithm, all documents are loaded, however, 31 documents are discharged because they contain less than *doc_keywords_size_min* keywords.

Table 1. Test 1 - clustering results and performance.

	Keygraph 1	Keygraph 2	Keygraph 3
Clustered documents	404 (16,8%)	290 (12%)	294 (12,3%)
Nodes	397	814	814
Edges	362	634	634
Communities	101	125	125
Document clusters			
Before merging	(99)	(97)	(97)
After merging	82	77	(77)
After link analysis and merging			69
Documents with shared links outside clusters	9	9	0
Performance measures			
Accuracy	81%	85%	87%
Precision	60%	75%	73%
Recall	65%	69%	73%

The number of clustered documents in Keygraph 2 approximately decreases a 28% with respect to Keygraph 1, while Keygraph 3 obtains a number of clustered documents similar to Keygraph 2. The first version is able to classify more documents than the other versions, but this does not mean that the first version is the best, because we must verify that the news in the cluster are about the same event.

In Fig. 4, we represent how many correlations were found among news: the first column represents correlations in the link cluster baseline (see Sect. 4.2 for the definition), the other columns show how many correlations are found by Keygraph 1, 2 and 3. The total number of correlations in link cluster is 75. 71 of them are among news on the same channel and only 4 among news on different channels. Nevertheless, the correlations found by Keygraph are many more than those found by using only link cluster. Keygraph 1 is able to find approximately the double of correlations than the other versions. In contrast to the results in link cluster, there is not a marked difference between same-channel correlations and different-channel correlations.

4.4   Test 2 - Multichannel Publishing by Different Newspapers

The second test has been executed on the news published by the 21 Italian newspapers on all channels. This test set contains 21457 news (5505 published

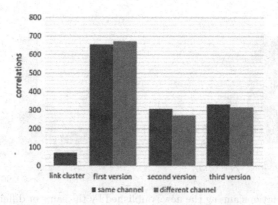

Fig. 4. Test 1 - the correlations among the published news.

on the Website -26%- 8867 on Facebook -41%-, and 7085 on Twitter -33%-) and 26063 links connecting the news in the test set is (so, on average, a news has more links). Besides, 3233 documents share at least one link with other document. All documents are loaded, but 281 are discharged because they have few keywords.

As shown in Table 2, the clustered documents in Keygraph 2 are less than the half of clustered documents in Keygraph 1, the same happens with the number of document clusters. In Keygraph 3 the clustered documents slightly increases w.r.t. Keygraph 2 and the document clusters decreases, like in the previous test.

In Fig. 5, the first columns show that in link cluster most of the correlations are between news published on the same channel or by the same newspaper. The other columns reveal that the Keygraph algorithm finds much more correlations between news: the first version finds many more correlations than the other two versions.

Table 2. Test 2 - clustering results and performance.

	Keygraph 1	Keygraph 2	Keygraph 3
Clustered documents	2777 (13%)	1025 (4,8%)	1074 (5%)
Nodes	4972	9916	9916
Edges	13575	63487	63487
Communities	789	761	761
Document clusters			
Before merging	(572)	(278)	(278)
After merging	514	265	(265)
After link analysis and merging			261
Documents with shared links outside clusters	135	52	0
Performance measures			
Accuracy	83%	88%	90%
Precision	75%	85%	86%
Recall	75%	79%	82%

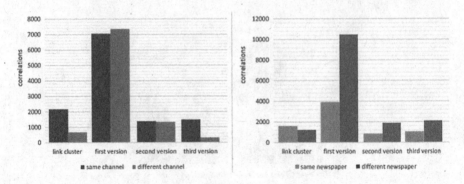

Fig. 5. The correlations among the news published by the same or different newspaper.

5   Conclusion

This paper represents an introductory analysis on the news published by the main Italian newspapers. By exploiting three extended versions of a graph analytical approach for topic detection and automatic indexing of documents, called Keygraph [9], we demonstrated how to cluster the news around the main topics to understand correlations and compare news published on different channels and different newspapers.

A preliminary evaluation of the three extended Keygraph versions on the news published in a 5 days period has shown promising results. Keygraph 3 was able to identify the main topics within the publications of a single newspaper reaching a 73% of precision and recall and also within the publications of 20 newspapers on several on-line channels reaching a 86% of precision and 82% of recall.

Future work will be focused on comparing the Keygraph algorithm w.r.t. other topic models such as LSA or LDA [4]. Moreover, we would like to investigate how disambiguation techniques might improved the results of Keygraph [3,10].

References

1. Allan, J. (ed.): Topic Detection and Tracking: Event-based Information Organization. Kluwer Academic Publishers, Norwell (2002)
2. Atefeh, F., Khreich, W.: A survey of techniques for event detection in Twitter. Comput. Intell. **31**(1), 132–164 (2015)
3. Bergamaschi, S., Beneventano, D., Po, L., Sorrentino, S.: Automatic normalization and annotation for discovering semantic mappings. In: Ceri, S., Brambilla, M. (eds.) Search Computing. LNCS, vol. 6585, pp. 85–100. Springer, Heidelberg (2011). doi:10.1007/978-3-642-19668-3_8
4. Bergamaschi, S., Po, L., Sorrentino, S.: Comparing topic models for a movie recommendation system. WEBIST **2**, 172–183 (2014)

5. Fiscus, J.G., Doddington, G.R.: Topic detection and tracking. In: Allan, J. (ed.) Topic Detection and Tracking Evaluation Overview, pp. 17–31. Kluwer Academic Publishers, Norwell (2002)
6. Garrido, A.L., Buey, M.G., Escudero, S., Ilarri, S., Mena, E., Silveira, S.B.: TM-gen: a topic map generator from text documents. In: 25th IEEE International Conference on Tools with Artificial Intelligence, Washington (USA). IEEE Computer Society, November 2013
7. Rajaraman, A., Ullman, J.D.: Mining of Massive Datasets. Cambridge University Press, New York (2011)
8. Sayyadi, H., Hurst, M., Maykov, A.: Event detection and tracking in social streams. In: Proceedings of the International Conference on Weblogs and Social Media (ICWSM 2009). AAAI (2009)
9. Sayyadi, H., Raschid, L.: A graph analytical approach for topic detection. ACM Trans. Internet Technol. **13**(2), 4:1–4:23 (2013)
10. Trillo, R., Po, L., Ilarri, S., Bergamaschi, S., Mena, E.: Using semantic techniques to access web data. Inf. Syst. **36**(2), 117–133 (2011)
11. Veglis, A.: Cross-media publishing by US newspapers. J. Electron. Publ. **10**(2), 131–150 (2007)

Random Walks Analysis on Graph Modelled Multimodal Collections

Serwah Sabetghadam, Mihai Lupu[✉], and Andreas Rauber

Institute of Software Technology and Interactive Systems,
Vienna University of Technology, Vienna, Austria
{sabetghadam,lupu,rauber}@ifs.tuwien.ac.at

Abstract. Nowadays, there is a proliferation of information objects from different modalities—Text, Image, Audio, Video. Different types of relations between information objects (e.g. similarity or semantic) has motivated graph-based search in multimodal Information Retrieval. In this paper, we formulate a Random Walks problem along our model for multimodal IR, that is robust over different distributions of modalities. We investigate query-dependent and query-independent Random Walks on our model. The results show that the query-dependent Random Walks provides higher precision value than query-independent Random Walks. We additionally investigate the contribution of the graph structure (quantified by the number and weights of incoming and outgoing links) to the final ranking in both types of Random Walks. We observed that query-dependent Random Walks is less dependent on the graph structure. The experiments are applied on a multimodal collection with about 400,000 documents and images.

1 Introduction

We observe a rapid growth in online data from different modalities, exacerbated by personal generated data. This trend creates the need for seamless information retrieval from different modalities such as video, audio, image, and text. Multimodal information retrieval is about the search for information of any modality on the web, with unimodal or multimodal queries. For instance a unimodal query may contain only keywords, whereas multimodal queries may be a combination of keywords, images, video clips or music files.

Multimodal IR presents a great challenge since it deals with several data types or modalities, each having its own intrinsic retrieval model. Research in this area has a wide range from associating image with text search scores, to sophisticated fusion of multiple modalities [4,6,11,12].

Fundamental IR is based on retrieval from independent documents. We refer to this type of IR as *non-structured IR*. Going beyond the document itself, in modern IR settings, documents are usually not isolated objects. Instead, they are frequently connected to other objects, via hyperlinks or meta-data [15]. They provide mutual information on each other—forming a background information model that may be used explicitly. Sometimes this information link is explicit

© Springer International Publishing AG 2017
A. Calì et al. (Eds.): IKC 2016, LNCS 10151, pp. 76–88, 2017.
DOI: 10.1007/978-3-319-53640-8_7

as related information (e.g. a Wikipedia document and its image), resulting in a network of related objects; sometimes it is inherent in the information object, e.g. similar pitch histogram of two music files. Moreover, user-generated multimodal content, domain-specific multimodal collections, or platforms like the semantic web impose *structured IR* based on links between different information objects. Since 2005 there is a trend towards leveraging both structured and non-structured IR [3,5,9] modelled in a graph of information objects.

Mostly, research in the area of structured IR creates a graph of one type of relations e.g., similarity links with one modality (images/videos) [7,24,25]. Others use only semantic relations between information objects based on various semantic databases [5,16,21]. For instance, Rocha et al. [16] model the knowledge-base of a sample website by adding semantic links between various entities of the website. However, the data today is multimodal. We proposed a model for multimodal IR [18], which supports heterogeneous links and information object.

One of the main challenges in such graphs in IR is to find relevant documents for a specific query. There are different approaches for graph traversal. One of the well-known methods used in IR is Random Walks. Usually the stationary distribution obtained by such Random Walks is leveraged to compute the final scores of documents and images after reranking [7,24,25]. However, it is mostly used in a graph with similarity links between one type of modality. In this paper, we investigate the role of query-dependent and query-independent Random Walks in final performance on a multimodal graph with different types of links. In query-dependent Random Walks we consider relevancy to the query in each step of traversal. In query-independent routing the only parameter influencing the traversal is the predefined weighting in the graph. We explore the role of the Metropolis-Hastings algorithm as query-dependent Random Walks on our graph. The reason to choose this method is that it can be an approach towards true relevance probability of the nodes to the query [20].

In order to gain extra insights, we track the score distribution from initial states in the graph until the convergence state. In earlier steps, Random Walks is biased towards initial text search scores. Further in the graph both initial search results and contextual relationships are considered. Additionally, we investigate the contribution of the graph structure (quantified by the number and weights of incoming and outgoing links) to the final ranking in both types of Random Walks.

The experiments are conducted on ImageCLEF 2011 Wikipedia collection. It is a multimodal collection with about 400,000 documents and images. The results show that with query-dependent Random Walks we obtain higher precision values. This demonstrates the effect of combining graph structure with query relevancy of different information objects in multimodal IR.

The paper is structured as follows: in the next section, we address the related work, followed in Sect. 3 by the basic definition of our model and hybrid search. The experiment design and results is shown in Sect. 4. Finally, conclusions and future work are presented in Sect. 4.5.

2 Related Work

Mei et al. [13] did a thorough survey on reranking methods in multimodal IR. They categorize related work in four groups: (1) Self-reranking: mining knowledge from initial ranked list. One of the methods is to use pseudo-relevance feedback. (2) Example-based reranking: leveraging user-provided query examples to detect relevant patterns. The examples along with the text query are used to combine the results. (3) Crowd-reranking: utilizing knowledge obtained from crowd as user-labeled data to perform meta-search in a supervised manner. (4) Interactive reranking which reranks involving user interactions.

Graph-based reranking method [7,8,10] is a subset of first category. Mostly the methods in this category are inspired by Page-Rank techniques for document search. In this way, the relevance score of a document is propagated through the entire graph structure. Usually a graph $G =< V, E >$ is made based on the initial ranked list, where $v \in V$ corresponds to a visual document and $e \in E$ is the similarity link between the two objects. The links between the images/videos are similarity links. The initial relevance scores of each document can be propagated to other similar nodes until the graph reaches a stationary distribution state.

Jing et al. [8] employ PageRank to rerank image search results. The hyperlinks between images are based on their visual similarities. They choose the "authority" nodes to answer the image queries. Yao et al. [25] make a similarity graph of images and aim to find authority nodes as results for image queries. Through this model, both visual content and textual information of the images is explored. Tonon et al. [22] using a hybrid search on Linked Open Data try to retrieve better results by exploring selected semantic links. As a desktop search engine, Beagle++ utilizes a combination of indexed and structured search [14]. In Astera we provide a hybrid search model that is not limited to work on RDF data.

Compared to current research in leveraging Random Walks in multimodal retrieval, Astera is different from two aspects: (1) Our model covers heterogeneity in both information objects and in the link types, while related work in this area deal with one type of relation such as semantic or similarity. Further, they include one type of modality, e.g., image or video. (2) Random Walks in literature has been applied mostly in query-independent routing. While in Astera, we provide the possibility to test query-dependent and independent routing. For this, we use Metropolis-Hastings algorithm which alters the edge weights in the graph based on the query. Provided with a reasonable estimation for the probability of relevance, this method can provide a better approximation to the true relevancy probability distribution.

3 Model Representation

We define a graph model $G = (V, E)$ to represent information objects and their relationships, together with a general framework for computing similarity. In this graph, V is the set of vertices (including data objects and their facets) and E

is the set of edges. Each object in this graph may have a number of facets. By facet we mean inherent information or property of an information object. For instance, an image object may have several facets (e.g. edge histogram, texture representation). We define four types of relations between the objects in the graph. The relations and their characteristics and weightings are discussed in detail in [18]. We briefly repeat them here for completeness of the presentation:

- Semantic (α): a relation between two information objects that are semantically related in the collection. E.g., a capital of a country or an album of a singer.
- Part-of (β): a specific type of semantic relation, indicating an object as part of another object, e.g. an image in a document.
- Similarity (γ): relation between objects with the same modality, e.g. between the same facets of two objects.
- Facet (δ): linking an object to its representation(s). It is a directed edge from facet to the object.

With the exception of semantic link, all other links are uni-directional. We proposed to leverage the combination of faceted search with graph search to find relevant objects [19]. Our hybrid ranking method consists of two steps: (1) In the first step, we perform an initial search with Lucene and/or LIRE to obtain a set of activation nodes, which is based on specific facet indexed results. (2) In the second step, pursued in this study, we use the initial result set of data objects (with normalized scores) as seeds and then exploit the graph structure and traverse it.

4 Experiments

4.1 Data Collection

We applied the ImageCLEF 2011 Wikipedia test collection. This collection is based on Wikipedia pages and their associated images. It is a multimodal collection, consisting of 125,828 documents and 237,434 images.

Each image in this collection has one metadata file that provides information about name, location, one or more associated parent documents in up to three languages (English, German and French), and textual image annotations (i.e. caption, description and comment). We generate a metadata document for each image out of its comment, caption and description. We created different relation types: the β relation between parent documents and images (as part of the document), δ relation between information objects and their facets. We use the 50 English query topics.

We enrich the collection by adding semantic and similarity links. For semantic links, we connect the ImageCLEF 2011 Wikipedia collection to correspondent DBPedia dump version. We add semantic links between two Wiki pages that there is a relation between their equivalent pages in DBpedia [17]. Similarity links (γ) connect the same facet type of two information objects. This could be theoretically between each two facets of the same type of two information objects in the collection. For example, it can be between color histogram facets of all images, or between textual TF.IDF facets of documents. For images, we consider their textual and visual facets for 10 most similar images/documents. The same for documents, we establish similarity links between a document and its 10 most similar neighbors.

4.2   Standard Text and Image Search

In the indexed search approach, as first phase of our hybrid search, we use Lucene indexing results both for documents and images. The computed scores in both modalities are normalized per topic between 0 and 1. Different indexings based on different facets are: (1) TF.IDF facet: We utilize default Lucene indexer, based on TF.IDF, as document facet. We index TF.IDF facet of image metadata as well. (2) CEDD facet: For image facets, we selected the Color and Edge Directivity Descriptor (CEDD) feature since it is considered the best method to extract purely visual results [1].

4.3   Graph Score Distribution

To start the experiments, first we want to obtain an understanding of score distribution in different steps in the graph. It usually takes more than 1000 steps that we reach to the stationary distribution state [23]. We use the term *step* as theoretical multiplication of the graph matrix. From an *iteration*, we mean each mathematical matrix multiplication, which may include one or many steps. We choose binary steps to create big blocks of matrix multiplication to recognize stationary distribution (convergence) in the graph.

To determine the number of steps needed, we check the convergence in each step. We calculate the difference of final score vector of previous iteration $a^{(t-1)}$ and this iteration $a^{(t)}$. We use absolute error $abs_{err} = |a^{(t)} - a^{(t-1)}|$ and relative error of $rel_{err} = abs_{err}/a^{(t-1)}$. Absolute error shows the difference of value of each node in the two vectors. Relative error value shows that how much we tolerate the absolute error of a node score in $a^{(t)}$ relatively to the score for the same node in $a^{(t-1)}$ vector.

Choosing the value for absolute error is based on the threshold of considering a value in the result vector for precision. We do not consider the score values less than $1e - 3$ in our ranked list. Therefore we chose $1e - 4$ as the absolute error value of convergence. We set the accuracy value of $5e - 2$ for relative error as enough accuracy between the two recent iterations. According to our checking for different topics, we reach to the convergence after step 2048 with 12 iterations.

We take snapshots from Random Walks steps for a sample topic (Topic 83). We show the score distribution in steps of $1, 2^3, 2^5, 2^7, 2^9$, and 2^{11} (Fig. 1). We observe that in the starting steps, fewer nodes have received the energy/score from starting points. As we traverse further in the graph, the energy is more evenly distributed. However, there are some nodes which have still higher energy than the others. Top images with higher scores are returned as ranked list. For

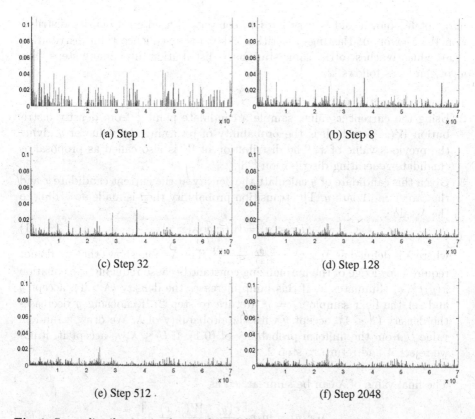

(a) Step 1

(b) Step 8

(c) Step 32

(d) Step 128

(e) Step 512

(f) Step 2048

Fig. 1. Score distribution in the graph in different steps for topic 83. The x axis is all the nodes in the graph which is 70,000. Units are in the scale of 10^4.

example for topic 83, the top ranked results obtain 0.3166, 0.1052, 0.0789, 0.0526 scores after 2048 steps.

After many iterations, we have the same amount of scores spread over the nodes. This score is the sum of normalized scores of the top results of a query. We observe in Fig. 1f that we have still nodes with distinguished higher score, which form the top results in the ranked list.

4.4 Query-Dependent and Query-Independent Routing

By query-independent routing, we mean that the traversal in each step is independent from the relevancy to a query. It is only based on the pre-defined weighting in the graph. We use Markov Chain Random Walks for this traversal. For query-dependent Random Walks we employ Metropolis-Hastings algorithm [2]. The reason to choose this method remains in the objective of this method. Metropolis-Hastings is one of the algorithms based on MCMC (Monte Carlo Markov Chain) to obtain samples from a complex probability distribution $\pi(x)$. Suppose that we know a $\tilde{\pi}(x)$ that $\pi(x) = \frac{\tilde{\pi}(x)}{K}$. The normalizing constant K

may not be known and is very hard to compute. Based on a density distribution W, Metropolis-Hastings algorithm generates a sequence from distribution $\widetilde{\pi}(x)$, which reaches to the same stationary distribution after many steps. The algorithm is as follows [2]:

1. Start with initial value x that $\widetilde{\pi}(x) > 0$.
2. Using the current x value, sample a candidate point y from density distribution $W(x, y)$, which is the probability of returning the value of y giving the previous value of x. The distribution of W is also called as proposal or candidate-generating distribution [23].
3. Given this candidate of y calculate the density at the current candidate x and the target candidate y. The transition probability then is made according to the λ value.

$$Pr(x, y) = W(x, y)\lambda(x, y) \tag{1}$$

where λ is defined as $\lambda(x, y) = \frac{\widetilde{\pi}(y)}{\widetilde{\pi}(x)} \cdot \frac{W(y,x)}{W(x,y)}$. The λ value shows that we do not require knowledge of the normalizing constant because reducing the equation $\widetilde{\pi}(y)/\widetilde{\pi}(x)$ eliminates it. If this jump increases the density ($\lambda > 1$), accept y and set the next sample $x_t = y$. Return to step 2. If choosing y decreases the density ($\lambda < 1$), accept y with the probability of λ. We draw a random value U from the uniform probability of (0,1). If $U \leq \lambda$ we accept it, if not we reject it and return to step 2.

The final value of λ can be summarized as

$$\lambda(x, y) = min\left[\frac{\widetilde{\pi}(y)}{\widetilde{\pi}(x)} \cdot \frac{W(y, x)}{W(x, y)}, 1\right] \tag{2}$$

Mapped to our problem, the proposed matrix W is our stochastic transition matrix of the graph. We satisfy the stochastic property by assigning the weight $1/N$ on each edge of a node with N neighbours. As a stationary distribution over the set of nodes, we would like to approach the *true* relevance probability distribution by the indexing ranked results. This is the $\pi(x)$ distribution from which we cannot directly sample. Instead, we have the $\widetilde{\pi}(x)$ which could be a relevance scoring function (e.g. a BM25 score between the information object x_i and the query). Metropolis-Hastings formally provides us with a method to sample from the probability distribution, if the approximate probability $\widetilde{\pi}(x)$ is properly chosen.

With Metropolis-Hastings, we provide a query-dependent traversal. In each step, we need the relevancy of each node to the query. We use BM25 method as $\widetilde{\pi}(x)$ to compute the relevancy of the nodes to the query. For this reason, we provide the relevancy of each node based on their facets to the query. In our previous analysis [20], we had better results for textual facets. Therefore, we choose document and image metadata TF.IDF facet for computing relevancy in each step. We consider all three languages of English, German and French for finding relevancy of an information object to the query.

From here, we compare these two methods, Random Walks and Metropolis-Hastings as query-independent and query-dependent Random Walks in our experiments.

Precision Analysis. We compare the performance of Random Walks and Metropolis-Hastings in this experiment. We calculate the precision and recall at cut-off for 25 topics with both methods. We choose TF.IDF facet of documents and image metadata for computing relevancy in each step. We start with top 20 results based on these facets for each query. We traverse the graph from these 40 starting points. We consider all three languages of English, German and French for finding relevancy of an information object to the query. Specially this is needed for image metadata indexes. Because many images do not contain metadata information in all languages. We consider the relevancy of an image in the three languages. This way, the relevancy value is influenced by the relations of that node.

The result in Table 1 shows that by leveraging Random Walks, in the steps higher than 16, the precision value is zero. This indicates that the top ranked results in stationary distribution are not relevant to the query. We are more interested in graph behaviour in higher steps than 16 with these two algorithms. We observe from Table 2 that with Metropolis-Hastings we obtain positive values for precision and recall in all iterations. Comparing the two tables, we observe lower values only in the first step for Metropolis-Hastings. The reason is that choosing the top images are influenced by their relevancy to the query. Our relevancy function considers only metadata indexes and all images do not have enough text in this part. However, from the 2nd step, Metropolis-Hastings keeps higher values in precision and recall.

In the stationary distribution with Random Walks, the only influencing parameter in the result vector is the graph structure. However, with Metropolis-Hastings, we include in addition the relevancy to the query of each information object in each step. The precision values could increase by using better relevancy functions for images and documents.

Correlation Analysis. In previous experiment on precision at cut-off, we found that in higher steps, the top ranked results are rarely relevant to the query. Why did these images obtain higher scores?

We perform rank correlation analysis based on the top ranked results and the number of their neighbours. We want to investigate the result bias to the number of nodes that a node is connected to. In each step, we get the top ranked images. For each image we find the number of its neighbours and create another ranked list. We calculate the correlation between these two lists. We use Spearman correlation, as it does not assume any assumption about the distribution of data. It is an appropriate choice when the variables are measured on an ordinal scale. We compute the correlation values in each step for both algorithms.

We observe from Fig. 2a that both methods show low correlation value between the top ranked results and the number of incoming links. However,

Table 1. Performance result with
Random Walks

iter	steps	p@10	r@10	p@20	r@20
1	1	0.2267	0.0815	0.1767	0.1111
2	2	0.1571	0.0478	0.15	0.0866
3	4	0.1	0.0293	0.1067	0.0523
4	8	0.0857	0.0209	0.0643	0.0364
5	16	0.0133	0.0027	0.0333	0.0175
6	32	0.0	0.0	0.0	0.0
7	64	0.0	0.0	0.0	0.0
8	128	0.0	0.0	0.0	0.0
9	256	0.0	0.0	0.0	0.0
10	512	0.0	0.0	0.0	0.0
11	1024	0.0	0.0	0.0	0.0
12	2048	0.0	0.0	0.0	0.0
13	4096	0.0	0.0	0.0	0.0

Table 2. Performance result with
Metropolis-Hastings

iter	steps	p@10	r@10	p@20	r@20
1	1	0.1958	0.0501	0.1479	0.0711
2	2	**0.216**	**0.065**	**0.16**	**0.0836**
3	4	0.175	0.0432	0.1562	0.0787
4	8	0.1333	0.029	0.1292	0.0582
5	16	0.1125	0.0284	0.0833	0.0354
5	32	0.068	0.0145	0.056	0.0189
7	64	0.032	0.0079	0.018	0.0081
8	128	**0.0042**	**0.0038**	**0.0021**	**0.0038**
9	256	0.0042	0.0038	0.0021	0.0038
10	512	0.0042	0.0038	0.0021	0.0038
11	1024	0.0042	0.0038	0.0021	0.0038
12	2048	0.0042	0.0038	0.0021	0.0038
13	4096	**0.0042**	**0.0038**	**0.0021**	**0.0038**

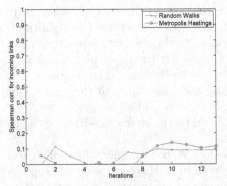

(a) Correlation between the top ranked results
and the number of **incoming links**

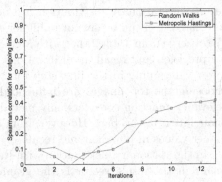

(b) Correlation between the top ranked results
and the number of **outgoing links**

Fig. 2. Correlation between top results and the links of these nodes

Metropolis-Hastings as a query-dependent traversal method, shows lower correlation values than Random Walks up to the step 2^9. We performed the same analysis with the number of outgoing links of top ranked results (Fig. 2b). We observe that the correlation between the rank and the number of outgoing links in both methods is less than 0.5. However, both algorithms show higher correlation value to the number of outgoing links. This shows that nodes with high fan-outs in the graph influence both algorithms.

Further, we investigate the rank correlation of top ranked images to the sum of the weights around a node. In Fig. 3a and b, we show the correlation result between the top ranked nodes and the sum of the weights on the incoming and outgoing edges of each top result node. We observe that Metropolis-Hastings

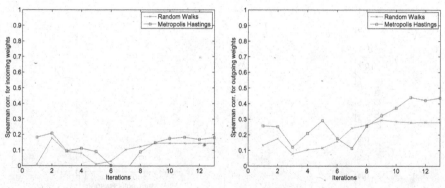

(a) Correlation between the top ranked results and the **incoming weights**

(b) Correlation between the top ranked results and the **outgoing weights**

Fig. 3. Correlation between top results and the weights of these nodes

holds higher value for correlation value in both cases. We find that this method is more influenced by the weighting in the graph. The reason is that one of the parameters to choose the next candidate in λ (Eq. 2) is the link weight. Therefore, the weights on incoming links of a node influences the jump decision to the next neighbour in Metropolis Hastings algorithm.

Score Memory. After correlation analysis, we want to investigate how fast the initial score influence is overriden by the graph structure using these methods. We compare two result vectors $(a_1^{(t)}, a_2^{(t)})$ in each iteration: one computed based on initial vector $(a_1^{(0)})$ composed of Lucene results; one computed based on equal score of $1/N$ for all nodes $(a_2^{(0)})$. This way, there is no preference between initiating points and we have only the effect of the graph structure. We compute the result vector in each step based on Equation

$$a^{(t)} = a^{(0)} \cdot W^t \tag{3}$$

where the $a^{(0)}$ vector is composed of top ranked nodes of query facets (as non-zero elements), and visited neighbours through traversal (as zero elements). The final vector, $a^{(t)}$, provides the final activation value of all nodes.

For each topic, in steps of power 2 (1, 2, 4,..., 4096), we calculate the $a_1^{(t)} = a_1^{(0)} \cdot W^t$ and $a_2^{(t)} = a_2^{(0)} \cdot W^t$ vectors. We calculate the Cosine similarity between $a_1^{(t)}$ and $a_2^{(t)}$ for both methods (Fig. 4). We observe that, in the first and second iterations in both methods, the similarity is very low—the results are highly dependent on the starting point scores. Surprisingly after step 2^6, we observe that the result vectors of both methods approach to highly similar values of $a_2^{(t)}$. This is exactly what the stationary distribution means that the stabilized values in each node is independent of initial values. However, we observe that with Metropolis-Hastings, we have lower similarity values in the initial steps.

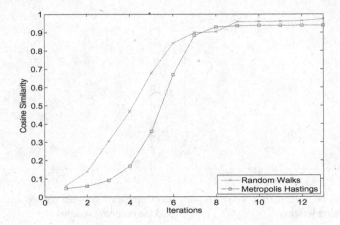

Fig. 4. The rate that the initial score is overriden by the graph structure with Random Walks and Metropolis-Hastings

This observation indicates that with Metropolis-Hastings, we can expect more relevant results in the first steps.

4.5 Conclusion

We formulated a Random Walks problem on our previously proposed model for graph-based multimodal IR. We have the opportunity to examine query dependent traversal, in which weights in the graph are affected by relevancy of source and target nodes to the query. We compared the performance of leveraging query-dependent and independent Random Walks in our graph model. The results show that by considering query relevancy in each step, we obtain higher precision results. This result can be improved by using better relevancy functions. We found that with Metropolis-Hastings as a query-dependent method, we can expect more relevant results not only in the first 32 steps.

We performed rank correlation analysis between the top ranked results and the number of incoming/outgoing links of these nodes. The results showed that with Metropolis Hastings we show less correlation value to the number of incoming links. Further, the correlation analysis results with the sum of the weights on incoming/outgoing links showed that weighting in the graph influences the Metropolis Hastings results.

This opens the challenge to address as the future work in this direction: (1) How to define the probability in a heterogeneous network with multi-modality and different link types. Satisfying the stochastic property in such a graph is a challenge ahead. How we define the probability in different relation types of part-of, similarity or semantic and meanwhile satisfying the stochastic property on the edges of a typical node is essential to evaluate the real effect of query-dependent Random Walks in our model. (2) How much expensive is this approach regarding the need of high number of transitions until the matrix burns in? (3) How to satisfy the stochastic property in a graph with heterogeneous nodes and relation types?

References

1. Berber, T., Vahid, A.H., Ozturkmenoglu, O., Hamed, R.G., Alpkocak, A.: Demir at imageclefwiki 2011: evaluating different weighting schemes in information retrieval. In: CLEF (2011)
2. Chib, S., Greenberg, E.: Understanding the metropolis-hastings algorithm. Am. Stat. **49**(4), 327–335 (1995)
3. Delbru, R., Toupikov, N., Catasta, M., Tummarello, G.: A node indexing scheme for web entity retrieval. In: Aroyo, L., Antoniou, G., Hyvönen, E., Teije, A., Stuckenschmidt, H., Cabral, L., Tudorache, T. (eds.) ESWC 2010. LNCS, vol. 6089, pp. 240–256. Springer, Heidelberg (2010). doi:10.1007/978-3-642-13489-0_17
4. Duan, L., Li, W., Tsang, I.W., Xu, D.: Improving web image search by bag-based reranking. IEEE Trans. Image Process. **20**(11), 3280–3290 (2011)
5. Elbassuoni, S., Blanco, R.: Keyword search over RDF graphs. In: CIKM (2011)
6. Fergus, R., Fei-Fei, L., Perona, P., Zisserman, A.: Learning object categories from google's image search. In: Proceedings of International Conference on Computer Vision (2005)
7. Hsu, W.H., Kennedy, L.S., Chang, S.-F.: Video search reranking through random walk over document-level context graph. In: MULTIMEDIA (2007)
8. Jing, Y., Baluja, S.: Visualrank: applying pagerank to large-scale image search. IEEE Trans. Pattern Anal. Mach. Intell. **30**(11), 1877–1890 (2008)
9. Kasneci, G., Suchanek, F., Ifrim, G., Ramanath, M., Weikum, G.: Naga: searching and ranking knowledge. In: ICDE (2008)
10. Liu, Y., Mei, T.: Optimizing visual search reranking via pairwise learning. IEEE Trans. Multimedia **13**(2), 280–291 (2011)
11. Martinet, J., Satoh, S.: An information theoretic approach for automatic document annotation from intermodal analysis. In: Workshop on Multimodal Information Retrieval (2007)
12. Donald, K.M., Smeaton, A.F.: A comparison of score, rank and probability-based fusion methods for video shot retrieval. In: Leow, W.-K., Lew, M.S., Chua, T.-S., Ma, W.-Y., Chaisorn, L., Bakker, E.M. (eds.) CIVR 2005. LNCS, vol. 3568, pp. 61–70. Springer, Heidelberg (2005). doi:10.1007/11526346_10
13. Mei, T., Rui, Y., Li, S., Tian, Q.: Multimedia search reranking: a literature survey. ACM Comput. Surv. (CSUR) **46**(3), 38 (2014)
14. Minack, E., Paiu, R., Costache, S., Demartini, G., Gaugaz, J., Ioannou, E., Chirita, P.-A., Nejdl, W.: Leveraging personal metadata for desktop search: the beagle++ system. J. Web Semant. Sci. Serv. Agents WWW **8**(1), 37–54 (2010)
15. Minkov, E., Cohen, W.W., Ng, A.Y.: Contextual search and name disambiguation in email using graphs. In: Proceedings of the 29th Annual International ACM SIGIR Conference on Research and Development in Information Retrieval, pp. 27–34 (2006)
16. Rocha, C., Schwabe, D., Aragao, M.P.: A hybrid approach for searching in the semantic web. In: WWW (2004)
17. Sabetghadam, S., Lupu, M., Rauber, A.: Astera - a generic model for multimodal information retrieval. In: Proceedings of Integrating IR Technologies for Professional Search Workshop (2013)

18. Sabetghadam, S., Lupu, M., Rauber, A.: A combined approach of structured and non-structured IR in multimodal domain. In: ICMR (2014)
19. Sabetghadam, S., Bierig, R., Rauber, A.: A hybrid approach for multi-faceted IR in multimodal domain. In: Kanoulas, E., Lupu, M., Clough, P., Sanderson, M., Hall, M., Hanbury, A., Toms, E. (eds.) CLEF 2014. LNCS, vol. 8685, pp. 86–97. Springer, Heidelberg (2014). doi:10.1007/978-3-319-11382-1_9
20. Sabetghadam, S., Lupu, M., Rauber, A.: Leveraging metropolis-hastings algorithm on graph-based model for multimodal IR. In: GSB 2015: First International Workshop on Graph Search and Beyond (2015)
21. Tiddi, I., dAquin, M., Motta, E.: Walking linked data: a graph traversal approach to explain clusters. In: Proceedings of the Fifth International Workshop on Consuming Linked Data, COLD (2014)
22. Tonon, A., Demartini, G., Cudré-Mauroux, P.: Combining inverted indices and structured search for ad-hoc object retrieval. In: SIGIR (2012)
23. Walsh, B.: Markov Chain Monte Carlo and Gibbs sampling (2004)
24. Wang, M., Li, H., Tao, D., Lu, K., Wu, X.: Multimodal graph-based reranking for web image search. IEEE Trans. Image Process. **21**(11), 4649–4661 (2012)
25. Yao, T., Mei, T., Ngo, C.-W.: Co-reranking by mutual reinforcement for image search. In: CIVR (2010)

Text and Digital Libraries

A Software Processing Chain
for Evaluating Thesaurus Quality

Javier Lacasta[1]([✉]), Gilles Falquet[2],
Javier Nogueras-Iso[1], and Javier Zarazaga-Soria[1]

[1] Computer Science and Systems Engineering Department,
Universidad de Zaragoza, Zaragoza, Spain
jlacasta@unizar.es
[2] Centre Universitaire D'Informatique, Université de Genève, Geneva, Switzerland

Abstract. Thesauri are knowledge models commonly used for information classification and retrieval whose structure is defined by standards that describe the main features the concepts and relations must have. However, following these standards requires a deep knowledge of the field the thesaurus is going to cover and experience in their creation. To help in this task, this paper describes a software processing chain that provides different validation components that evaluates the quality of the main thesaurus features.

Keywords: Thesaurus · Digital libraries · Information retrieval · Thesaurus quality · Ontology alignment

1 Introduction

The resources in metadata repositories are frequently classified using thesauri because of their simple structure, the established standards [7] and the integrated support provided by most catalog tools. Keyword based search is the standard for performing searches in many information systems and thesauri are one of the most used models to organize and relate the keywords [19]. The construction of a thesaurus requires a careful selection of the concepts in an area of knowledge and their interrelations in an appropriate general-to-specific hierarchy [5]. However, many factors, such as the lack of experience, costs savings, or the over-adaptation to a data collection, produce models with heterogeneous concepts and relations [4].

Simple edition tools for thesauri focus on providing a suitable environment for the creators to define concepts and relations in a collaborative way. In these tools, quality control is mainly focused on the definition of a human-oriented process were an editor reviews the work previous to its final inclusion in the thesaurus. This approach is especially valid for small thesaurus in which a person can maintain the control over the entire model. However, as the thesaurus size grows, this process becomes more difficult and problems of terminological

© Springer International Publishing AG 2017
A. Calì et al. (Eds.): IKC 2016, LNCS 10151, pp. 91–99, 2017.
DOI: 10.1007/978-3-319-53640-8_8

heterogeneity, overload of specificity, lexical issues in concept labels or unclear hierarchies become common [4,16].

In a previous work [10], we described a process to detect issues in the thesaurus according to ISO 25964 specification [7] at all different levels (labels, concepts, and relations). This paper continues in this line of work, and describes the software processing chain created to implement the validation tasks, and the integration framework used to merge these components into a complete validation tool. Instead of a monolithic analysis process, we have opted for using a modular approach in which each library component analyses a single feature. This greatly increases the flexibility of use of the library and facilitates its use in contexts where not all the thesaurus features are required. For example, to perform real-time validation of property values when they are defined in a thesaurus edition tool. Additionally, since some of the validation tasks are intensive in processing, the validation time can be quite considerable. Therefore, in addition to a modular approach, we have used a framework that greatly simplifies the parallel execution of the validation tasks in a single machine or cluster.

The paper is structured as follows. Section 2 summarizes the quality features for which we have created validation components, and reviews existent software solutions. Section 3 describes these components and how they are integrated in a tool. The paper ends with some conclusions and an outlook on future work.

2   Thesaurus Quality Features and Existent Quality Analysis Tools

"The quality" is a measure of excellence or a state of being free from defects, deficiencies and significant variations. ISO 8402 [8] defines the quality as "the totality of features and characteristics of a product or service that bears its ability to satisfy stated or implied needs".

The main sources to identify the quality features of a thesaurus are the existing construction guidelines. They range from practice manuals such as Aitchison et al. [1], to the current international standard ISO 25964 [7]. Pinto [13], Kless and Milton [9], and Mader and Haslhofer [11] are the principal studies focused on identifying the features that determine the quality of thesauri. They focus on concepts, terms, structure and documentation parts, and describe features that need to be reviewed at each level. This information has been compiled from specifications, previous works in the area and user surveys.

From the features described in these works, we have developed components to automatically analyze the main elements described in ISO 25964:

Property completeness measures: These measures are focused on the identification of lacking properties. We analyse the completeness and uniqueness of preferred labels and completeness of definitions.

Property content measures: Their objective is to locate invalid values inside labels. We focus on detecting non-alphabetic characters, adverbs, initial articles, and acronyms (in preferred labels).

Property context measures: These are focused on identifying anomalies involving several labels. This includes detecting duplicated labels and inconsistencies in the use of uppercase and plurals.

Property complexity measures: They provide a measure of the syntactic complexity of the labels, in terms of the use of prepositions, conjunctions and adjectives.

Relation coherence measures: They indicate if the relations are complete, coherent, and semantically correct. RT (Related Term) analysis focuses on detecting non-informative relations (they link hierarchically related concepts). BT/NT (Broader/Narrower Terms) analysis searches for cycles in the model, unlinked concepts and relations that do not associate a superordinate with a subordinate concept. According to ISO 25964, the superordinate must represent a class or whole and subordinate its members or parts.

There are some validation tools such as Mader and Haslhofer [11], Suominen and Mader [17], Eckert [3] or Poveda-Villalón et al. [14] that analyze some of these features, but not all of them. They focus on reviewing the completeness of the properties and relations, and existence of cycles. Only Eckert [3] provides a method to detect issues in the relations, but at the cost of using a collection classified with the thesaurus. Our work deals with all of them, this includes those related to the syntaxis and semantics of the labels, concepts and relations (e.g., the use of adverbs, acronyms, and the meaning of BT/NT relations). The developed components analyze the thesaurus features through structural, lexical, syntactical and semantical checks that validate the labels describing the concepts and the provided relations.

3 Design of the Software Processing Chain

For the development of the software that performs the validation of the previously described features, we have opted for a modular approach in which each feature is reviewed by a different class in the validation library. This provides a great flexibility in terms of adjusting the system to different needs, such as the development of tools that only perform a subset of the implemented validation tasks, and the addition or improvement of functionality.

To implement these modules we have used the Spring framework[1]. This framework simplifies the use of the dependency-injection pattern, allowing the definition of data flows between completely decoupled components through a configuration file (or even class annotations). These data flows can be defined so the unrelated parts can be automatically executed in parallel by different processes in the same or in different computers without the programmer having to program the distribution and aggregation of the data, or the synchronization of the processes. Additionally, it provides other useful functionality such as transaction control to deal with errors, and logging (between many others).

---

[1] https://spring.io/.

Some of the required validation tasks are quite intensive in processing due to their complexity and/or the amount of data the thesaurus contains. This makes difficult the construction of a fast validation tool that quickly performs the analysis. In this context, the use of Spring has facilitated us the construction of a parallel execution flow for unrelated validation tasks. Additionally, some of these tasks are composed of independent sub-processes, which are also independent between them. From a technical point of view, the validation tasks can be classified in three families: those that require the processing of some properties distributed along the whole thesaurus, those that require the analysis of several properties of each concept, and those that analyze the content of the instances of a single property. From them, only those analysis affecting the whole set of properties need to be executed as a whole. Those affecting a concept or a property can be divided in as many concepts or properties the thesaurus contains.

From a very general point of view, the validation tool is composed of a reader, a validator, and a report generator (see Fig. 1). The reader provides the access to the repository and loads the thesaurus in memory. The validator provides the data flow that decomposes the thesaurus and provides them to the different validation sub-components. Finally, the report generator creates a human friendly document describing the detected issues. The current implementation of the reader is focused on loading thesaurus in SKOS format [12] and the report generator provides a simple textual report. However, thanks to the development framework used, these components can be replaced by other ones able to load and generate other formats (e.g., HTML, Excel, PDF, and so on) with a simple modification of the process configuration file.

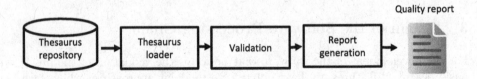

Fig. 1. Thesaurus validation process

The subset of the validation tasks that require the analysis of the whole thesaurus are the detection of non "Informative RT" relations, "BT/NT cycles" and "duplicated labels" (see Fig. 2). RT analysis involves the processing of all the other concepts in the same branch to detect if the are already related by a BT/NT relation. Cycles are located using a modified version of Tarjan's strongly connected components algorithm [18] that identifies the relation that generates the cycle (it points out to a broader concept). With respect to duplicated labels, it is needed to compare each label with the rest, but it has to be done aggregating them per language, since different languages can use the same word to describe a concept. Therefore, the validation task has been implemented to be independent of the language, and the framework has been configured to call it as many times as languages are in the thesaurus.

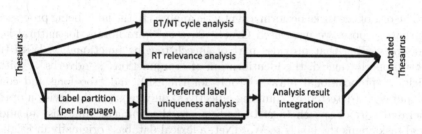

Fig. 2. Thesaurus level validation tasks

The validation tasks centered in detecting properties whose context is a single thesaurus concept are the detection of the completeness in the "definitions", "preferred labels" (there must be only one per concept and language) and "BT/NT relations" (no orphan concepts or branches). The data flow is depicted in Fig. 3. The algorithms used for these tasks are unremarkable, as they only check the existence of the properties and make annotations to the concept that are latter used in the report generation step.

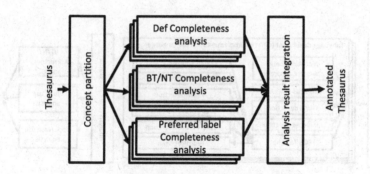

Fig. 3. Concept level validation tasks

The set of tasks focused on properties of a single label (preferred or alternative ones) includes some trivial and some complex tasks (see Fig. 4). In any case, an important characteristic all of them have is their dependence on the language. Therefore, is needed to define a new family of processors adapted to each language that implement all the tasks in this category.

Among the trivial tasks we found the identification of "non-alphabetic characters", "acronyms", "initial uppercase", and "plurals". The first three only require simple character comparisons, while plural detection has required the use of an adapted version of the Solr minimal stemmer [15] that detects plurals instead of removing them. The use of uppercase and plural in a label is not per se a quality feature, but the homogeneous use along all the thesaurus is. Therefore, in this phase, uppercase and plural labels are tagged so the real quality features can be analyzed latter in a simpler way.

The rest of the tasks perform a syntactic analysis to the label being processed. For that purpose, we have used GATE [2], a software library for natural language processing that provides part of speech tagging functionality. With the labels properly tagged, the identification of "conjunctions", "adverbs", "initial articles", "prepositional phrases" (only for English), and "too long and complex phrases" (detected by counting adjectives) only requires the revision of the generated part of speech tags. In addition to the syntactic analysis, an additional task aligns the labels to WordNet, a lexical database originally in English that groups nouns, verbs, adjectives and adverbs into sets of cognitive synonyms (Synsets), each expressing a distinct concept. Synsets are interlinked by means of conceptual-semantic and lexical relations providing a hypernym/hyponym hierarchy of semantically related concepts. This is done as a previous step for the identification of the semantic correctness of BT/NT relations. Since the thesaurus labels can be in languages different from English, instead of the pure WordNet, we have used the Open Multilingual WordNet (OMWordNet)[2]. This is an extension of WordNet that maintains the concept relation structure but incorporating labels in several additional languages. In this step, a direct lexical alignment is performed, annotating the label with all the WordNet concepts that share the label (ignoring case and plural).

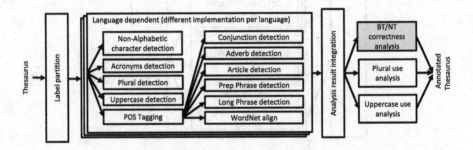

Fig. 4. Label level validation tasks

Once the plural, initial uppercase tags, and WordNet senses are added at each label, a last set of validation tasks that require this information as prerequisite is executed. These tasks are the check of a "homogeneous use of plurals and initial uppercase in labels", and the "semantic correctness of BT/NT relations". Plural and uppercase check just requires counting the occurrences of each feature and marking as incorrect those that are different from the majority. With respect to the detection of the correctness of BT/NT relations, it aligns each concept with the DOLCE ontology [6] to identify the semantic meaning of the relations and therefore identify the incorrect ones (see Fig. 5). DOLCE provides top level categories of concepts with a deep semantic net of relations between them. Some

---

[2] http://compling.hss.ntu.edu.sg/omw/.

of these relations, such as "participant" or "exact location of" intrinsically provide a superordinate and subordinate meaning and they can be considered as BT/NT specializations. These concepts are too generic, so a direct alignment with the thesaurus is not possible. Therefore, the alignment with DOLCE is done thanks to the previously obtained alignment with WordNet and a manual alignment between WordNet Synsets and DOLCE categories [6]. However, to be able to use the alignment, it is needed to select the correct meaning of each thesaurus in WordNet from the multiple ones provided by each language. This is done by selecting the most common Synset between the obtained for all the labels in different languages of a concept. Additionally, in the case where there are several alternatives with the same occurrences, the process uses previously established alignments of other concepts in the branch as the disambiguation context. In this case, the semantically closest sense in WordNet to those already selected for other concepts is the chosen one.

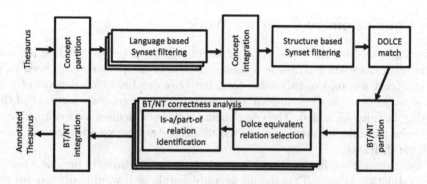

Fig. 5. Detail of the BT/NT semantic correctness validation task

The final step in this task reviews the concepts involved in each BT/NT relation to obtain the equivalent relation in DOLCE. We have identified three families of DOLCE relations (subclass, participation and location) that are compatible with BT/NT semantics. Therefore, when one relation in this family is found, the original BT/NT is tagged as correct, in other case is considered incorrect. A detailed description of each family is described next:

- The subclass relation indicates that the original concepts belong to hierarchically related categories. This does not ensure that the original BT/NT is correct, but because the thesaurus objective is to create generic to specific models, it is a good clue in that direction.
- The participation relation holds between perdurants (activities) and endurants (objects). It indicates elements that are part of an activity, which is a valid BT/NT meaning (e.g., horse piece is *part-of* a chess game or a car is *part-of* a car accident).
- With respect to the location related properties (from spatial to conceptual location), they may provide a *part-of* meaning (e.g., fountain *part-of* park)

or an *is-a* meaning (e.g., linear town *is-a* kind of urban morphology), both valid in the thesaurus context. If a BT/NT relation is assigned to one of these families, we consider it as correct.

The results of all these tasks are provided to the report generator that generates a textual file with a summary of the issues found. The quality of most of the properties is measured with the percentage of correctness of each of the analyzed features. This percentage is calculated in base to the number of properties/concepts/relations analyzed in a feature and the number of errors identified. Only the cycles in the thesaurus are counted with absolute numbers, as they are critical errors in the definition of the thesaurus where a percentage measure of correctness has no sense. It additionally generates additional log files with the list of errors identified. These logs are adapted to each analyzed feature, and identify the concept(s) involved in the erroneous element, and the value of the erroneous property (or relation, depending on the case).

4 Conclusions and Future Work

This paper has described the library components developed to validate the correctness of the main features of a thesaurus. These components have been used to construct a complete validation tool, but they can be used independently or arranged in other aggregation ways to analyse in other contexts a subset of the quality features identified. The components have been defined as decoupled as it has been possible, in order to allow the tool to be easily extended, reconfigured and parallelized.

Future work will be devoted to improve the current system to provide a public web validation system. This should be quite simple as it would only require the replacement of the components in charge of the thesaurus load and the validation report generation. Additionally, we want to use the validation components in a separate way to integrate them in a thesaurus edition tool, so that the thesaurus features can be validated as soon as a new element is added to the thesaurus. Finally, we want to continue to extend the validation components so that they can be used in other contexts apart from thesaurus analysis. For example, we would like to use our proposed processing chain for analyzing the quality of properties and relations in ontologies.

Acknowledgements. This work has been partially supported by the Keystone COST Action IC1302 and by the University of Zaragoza (project UZ2016-TEC-05).

References

1. Aitchison, J., Bawden, D., Gilchrist, A.: Thesaurus Construction and Use: A Practical Manual, Routledge (2000)
2. Cunningham, H., Maynard, D., Bontcheva, K., Tablan, V.: GATE: an architecture for development of robust HLT applications. In: Proceedings of the 40th annual meeting on association for computational linguistics, pp. 168–175. Association for Computational Linguistics (2002)

3. Eckert, K.: Usage-driven maintenance of knowledge organization systems. Ph.D. thesis, Universitat Mannheim (2012)
4. Fischer, D.H.: From thesauri towards ontologies? In: Structures and relations in knowledge organization - 5th International ISKO Conference, pp. 18–30, Lille, France, August 1998
5. Frakes, W.B., Baeza-Yates, R.: Thesaurus construction. In: Frakes, W.B., Baeza-Yates, R. (eds.) Information Retrieval: Data Structures & Algorithms, pp. 161–218. Addison Wesley, Reading (1992)
6. Gangemi, A., Guarino, N., Masolo, C., Oltramari, A.: Sweetening WORDNET with DOLCE. AI Mag. **24**(3), 13–24 (2003)
7. International Organization for Standardization: Thesauri and interoperability with other vocabularies. ISO 25694, International Organization for Standardization (ISO) (2011)
8. International Organization for Standarization: Quality management and quality assurance. ISO 8402, International Organization for Standarization (1994)
9. Kless, D., Milton, S.: Towards quality measures for evaluating thesauri. In: Sánchez-Alonso, S., Athanasiadis, I.N. (eds.) MTSR 2010. CCIS, vol. 108, pp. 312–319. Springer, Heidelberg (2010). doi:10.1007/978-3-642-16552-8_28
10. Lacasta, J., Falquet, G., Zarazaga-Soria, F.J., Nogueras-Iso, J.: An automatic method for reporting the quality of thesauri. Data Knowl. Eng. **104**, 1–14 (2016)
11. Mader, C., Haslhofer, B.: Perception and relevance of quality issues in web vocabularies. In: I-SEMANTICS 2013 Proceedings of the 9th International Conference on Semantic Systems (2013)
12. Miles, A., Bechhofer, S.: SKOS Simple Knowledge Organization System Reference. No. January in W3C Candidate Recommendation, W3C (2009)
13. Pinto, M.: A user view of the factors affecting quality of thesauri in social science databases. Libr. Inf. Sci. Res. **30**(3), 216–221 (2008)
14. Poveda-Villalón, M., Suárez-Figueroa, M.C., Gómez-Pérez, A.: Validating ontologies with OOPS!. In: Teije, A., Völker, J., Handschuh, S., Stuckenschmidt, H., d'Acquin, M., Nikolov, A., Aussenac-Gilles, N., Hernandez, N. (eds.) EKAW 2012. LNCS (LNAI), vol. 7603, pp. 267–281. Springer, Heidelberg (2012). doi:10.1007/978-3-642-33876-2_24
15. Savoy, J.: Report on CLEF-2001 experiments. Technical report, Institut interfacultaire d'informatique, Université de Neuchtel, Switzerland (2001)
16. Soergel, D.: Indexing Languages and Thesauri: Construction and Maintenance. Melville Pub, Company (1974)
17. Suominen, O., Mader, C.: Assessing and improving the quality of SKOS vocabularies. J. Data Semant. **3**(1), 47–73 (2014)
18. Tarjan, R.E.: Depth-first search and linear graph algorithms. SIAM J. Comput. **1**(2), 146–160 (1972)
19. Wielemaker, J., Hildebrand, M., Ossenbruggen, J., Schreiber, G.: Thesaurus-based search in large heterogeneous collections. In: Sheth, A., Staab, S., Dean, M., Paolucci, M., Maynard, D., Finin, T., Thirunarayan, K. (eds.) ISWC 2008. LNCS, vol. 5318, pp. 695–708. Springer, Heidelberg (2008). doi:10.1007/978-3-540-88564-1_44

Comparison of Collaborative and Content-Based Automatic Recommendation Approaches in a Digital Library of Serbian PhD Dissertations

Joel Azzopardi[1(✉)], Dragan Ivanovic[2], and Georgia Kapitsaki[3]

[1] University of Malta, Msida, Malta
joel.azzopardi@um.edu.mt
[2] University of Novi Sad, Novi Sad, Serbia
chenejac@uns.ac.rs
[3] University of Cyprus, Nicosia, Cyprus
gkapi@cs.ucy.ac.cy

Abstract. Digital libraries have become an excellent information resource for researchers. However, users of digital libraries would be served better by having the relevant items 'pushed' to them. In this research, we present various automatic recommendation systems to be used in a digital library of Serbian PhD Dissertations. We experiment with the use of Latent Semantic Analysis (LSA) in both content and collaborative recommendation approaches, and evaluate the use of different similarity functions. We find that the best results are obtained when using a collaborative approach that utilises LSA and Pearson similarity.

1 Introduction

The growth of the digital age has led to huge amounts of information being made available to users world wide through the World Wide Web. In fact, digital libraries have now become an essential tool for researchers. However, the sheer amount of information available to users precipitate the need for automated systems to filter the information to users according to their specified needs. Practically all digital libraries present a search interface whereby users can search for relevant content by expressing their needs in the form of queries. However, users may not even know exactly what they are looking for, and most often find it difficult to articulate their information needs through search queries [1,7,13].

Notwithstanding these difficulties, users typically find it easy to identify those documents that satisfy their information needs [5]. Automatic recommendation systems exploit this fact by constructing user models based on the users' prior behaviour, and then use these models to provide recommendations [2]. Apart from presenting 'new' (previously unknown) items of interest to users, automatic recommendation systems may also be used to predict an interest value for a specific item, or provide constrained recommendations from within a definite set of items (e.g. re-ranking search results thus providing personalised search) [20].

Automatic recommendation techniques can be classified into two main types [4,7,11,16,20], namely:

© Springer International Publishing AG 2017
A. Calì et al. (Eds.): IKC 2016, LNCS 10151, pp. 100–111, 2017.
DOI: 10.1007/978-3-319-53640-8_9

- **content-based techniques** – whereby recommendation is performed on the basis of similarity between the documents' contents; and
- **collaborative techniques** – whereby recommendation is performed on the basis of what other 'similar' users have found useful.

In this research, we present and compare a number of different content-based and collaborative techniques that were implemented to perform personalised recommendation on a digital library of PhD theses in Serbian. The University of Novi Sad has been maintaining a Digital library of PhD theses (PHD UNS) since 2012[1]. This library contains the full text of the dissertations, along with some other meta-data (that is not currently used in this recommendation system). Currently, PHD UNS users can search the digital library and manually browse the list of new PhD dissertations added to PHD UNS.

We also investigate the use of Latent Semantic Analysis (LSA) – originally proposed for Information Retrieval in [8] – in both content-based and collaborative algorithms. We also analyse the effect of different similarity functions.

The rest of this paper is organised as follows: Sect. 2 contains an overview of related research in this field; the system setup and the different recommendation algorithms implemented are described in Sect. 3; Sect. 4 describes the evaluation performed to determine the best-performing algorithms; and Sect. 5 discusses the main conclusions obtained and the future work planned.

2   Related Work

The typical processes within an automatic recommendation are [1,2,4,5,17–19]:

- **Feature Extraction** and **Logical Representation** of the information;
- **Filtering** – identification of relevant information; and
- **User feedback** to update the corresponding user models.

In content-based recommendation systems, the most common information representation scheme used is the *Bag-of-Words* representation [1,3–5,11,16–19]. According to Belkin [3], such simple word-based representations when used with the appropriate retrieval models, are effective, efficient and simple to implement. The simplicity of such representations implies that their use is feasible even in cases of very high volume of incoming information [3]. In the *Bag-of-Words* representation, both the documents and the user profiles are represented as weighted vectors [1,4,5,17–19]. Usually, vectors are weighted using *TF.IDF* [1,5,11,17–19]. Document terms are generally stemmed [1,3,17–19], and stop words are eliminated prior to the construction of the term vectors [1,3,17–19].

One short-coming of the *Bag-of-Words* representation is the assumption of term independence – this is a reasonable but not very realistic assumption [9]. As a solution, Foltz [9] extends the *Bag-of-Words* approach using *Latent Semantic Analysis* – a word-by-document matrix is constructed and decomposed into a set of orthogonal factors, and documents (and user profiles) are represented as

---

[1] http://dosird.uns.ac.rs/phd-uns-digital-library-phd-dissertations.

continuous values on each of the orthogonal index dimensions rather than as vectors of independent words.

Since collaborative recommendation systems calculate two items' degree of similarity based on the overlap of users that have rated both items positively, these items may be represented as vectors of user ratings [20]. Thus, collaborative recommendation may be simplified to the construction of a user-by-item matrix where each cell is the user's rating for a particular item.

After the documents/items have been represented logically, filtering is performed by comparing the items' logical representations with the user profiles. Recommendation is then performed on the basis of this similarity.

In content-based recommendation, filtering can be performed in two ways – **semantically** or **syntactically** [14]. One of the simplest syntactic methods used is keyword matching [9], where comparison between user profiles and documents can be performed by calculating the cosine similarity between the corresponding term vectors [11,15,17–19]. A document can then be classified as relevant or interesting for a user if the similarity to the user profile vector exceeds a predefined threshold [18,19]. However, as Foltz [9] points out, filtering based on keyword matching only has various issues – e.g. the words used may have multiple meanings, and the same concept can be described using different words [9].

Collaborative recommendation is performed by calculating the overlap between the different items, or the different users, in the user-item rating matrix [20]. Recommendations may be calculated using probabilistic (e.g. using Bayesian networks), and non-probabilistic methods. One of the most common non-probabilistic algorithms used is the *k-Nearest Neighbours* (*KNN*). [20] states that there is no published evidence of Bayesian networks performing better than KNN. KNN can be applied in its standard form, or else having it adjust the neighbours' vote weighting according to similarity (i.e. a more similar neighbour will have a higher voting weight).

Since recommendation systems calculate recommendations based on users' interests, their effectiveness depends on the use of appropriate user profiles and user feedback mechanisms. The user profile represents a stable (long-term) and structured information need of the corresponding user [15]. This may be contrasted to the user's short-term information need when using IR systems.

User-feedback mechanisms are divided into two, namely: **active** or **explicit feedback mechanisms** – where users specify their feedback directly (such as rating articles according to their interest); and **passive** or **implicit feedback mechanisms** – where feedback is collected indirectly from the user (for example by measuring the time spent by a user reading an article) [4,11]. *Active feedback* is considered to be more accurate [7,11]. For example, a system, where a user's click on an article is considered as a positive vote for that article, is much more noisy than prompting the user to rate the article on a scale between 1 and 5 [7]. On the other hand, *active feedback* entails extra effort from the user [11,14].

Content-based and Collaborative recommendation systems have their own advantages and disadvantages when compared to each other. Collaborative recommendation systems are generally considered more simple and straight-forward to implement [7]. Another advantage of collaborative recommendation is its domain independence – it does not require any content features to be extracted, and can be adapted to work in any domain and language with minimal effort [4,7,20]. Also, collaborative systems may lead to more unexpected recommendations [20]. On the other hand, content-based systems do not suffer from the user-sparsity problem – collaborative systems require a large number of users and an overlap of ratings for them to be effective and would not be feasible in a limited user environment [4,6,17,19,20].

A recommendation system can calculate recommendations using multiple techniques simultaneously – Pon [18] performs filtering using an *ensemble* of ranking functions. Albayrak [1] also has multiple filtering agents that employ different filtering techniques. In such ensemble systems, the weakness of individual techniques can be compensated by the strengths of others [1].

Moreover, recommendation systems can incorporate features from both content-based and collaborative recommendation techniques. Choi, in [6], reports improved results when predicting movie ratings by considering the content type of each item in addition to the different users' ratings for that item. Similarly, in [12], user ratings are combined with features extracted from review texts that accompany these ratings. Such combination helps in countering the sparse-matrix problem since by leveraging review texts, the system described in [12] is able to suggest previously unrated items with confidence, thus leading to a significant reported increase in prediction accuracy.

Notwithstanding the previous reported increases in effectiveness of hybrid systems, Garcin *et al.* [10] report that during experiments on personalised news recommendation, content-based and hybrid (combined content and collaborative techniques) approaches perform worse than pure collaborative techniques.

In comparison to these other works, which combine content-based with collaborative techniques, we apply and compare collaborative and content-based approaches to provide recommendations within digital libraries. To date, we have not yet applied hybrid approaches that combine both collaborative and content-based recommendation techniques.

3   Implementation

This section describes the recommendation system that was developed as part of this research to operate on the PHD UNS Digital Library. One challenge was that practically all the documents in this digital library are in Serbian. Our techniques were implemented to be language-independent.

3.1   The PHD UNS Digital Library

The PHD UNS digital library currently consists of 1897 PhD dissertations. Apart from very few exceptions, the dissertations are all in Serbian. A number of these dissertations are written using the Cyrillic alphabet, whilst others are written using the Latin alphabet[2].

Given that the dissertations' contents needed to be analysed to generate content-based recommendations, we developed a tool that performs the required pre-processing. The dissertations are converted from PDF to normal text using *pdftotext* utility that is commonly found in standard Linux distributions. The developed tool then performs the following tasks: conversion of the Cyrillic characters into their corresponding characters from the Latin alphabet; case folding, stop word removal and stemming; and conversion of non-ASCII characters to their corresponding characters from the ASCII character set. This pre-processing tool is the only language-dependent feature in our system.

Whilst the PHD UNS digital library has been active since 2012, the decision to start extending it to issue automatic recommendations was taken very recently. Hence, the software functionality to log the users' interactions in terms of downloads from this library was developed quite recently, so only recent download logs are available. Two sets of download logs were available at the time of this research – the first set covered the period from 20/2/2016 13:14:26 till 1/4/2016 08:50:53, whilst the second set covered the period between 2/4/2016 18:47:18 and 8/4/2016 10:49:59. The first set was used to train the system, whilst the second set was used for evaluation.

The download logs are initially pre-processed to remove logs generated by search engine bots. This is performed using a hand-crafted list of known bots and their corresponding "user agent" HTTP header fields. This filtering is performed since search engine bots are not real users, and can introduce noise that affect the recommender system's effectiveness.

In both the content-based and also in the collaborative recommendation systems, we consider the fact that a user has downloaded a dissertation as an implicit positive rating from that user for that dissertation. We do not consider any negative feedback.

3.2   Content-Based Recommendation

Prior to applying the content-based recommendation algorithms, a term-by-document matrix is constructed where the rows correspond to the different stemmed terms, and the columns represent the different dissertations. Each cell value contains the TF.IDF weight of the corresponding term within the corresponding document. This term-by-document matrix is also decomposed using Singular Value Decomposition (SVD) (for eventual use during the application of LSA).

In the non-LSA version of the content-based recommendation system, a user profile vector is built for each user by calculating the average vector from the vectors of the documents that the user has read, using Eq. 1 below:

---

[2] The Serbian language has a Latin and a Cyrillic alphabet, and the two alphabets are interchangeable.

$$\overrightarrow{UP(u)} = \frac{\sum\limits_{\vec{d} \in \overrightarrow{D_u}} \vec{d}}{|\overrightarrow{D_u}|} \tag{1}$$

where $\overrightarrow{UP(u)}$ refers to the user profile of user u and $\overrightarrow{D_u}$ represents the set of vectors for the documents that have been rated positively by user u.

The generated user profile vector $\overrightarrow{UP(u)}$ is then compared to the unread documents (i.e. documents that are not in $\overrightarrow{D_u}$) using the Cosine Similarity, and the list of recommendations is generated as a list of unread documents sorted according to similarity to $\overrightarrow{UP(u)}$.

For the LSA version of content-based recommendation system, the decomposed term-by-document matrix is used, with only a sub-set of the original dimensions, to construct the user profile vectors and then generate the list of recommendations as in the non-LSA system. The application of LSA is based on the premise that latent semantic features present in the original documents may be utilised for the benefit of the recommendation.

3.3   Collaborative Recommendation

As pre-processing to the collaborative recommendation systems, a user-by-item ratings matrix is constructed from the filtered download logs. The value of each cell is the number of downloads a user has performed for a particular dissertation. One should note that the PHD UNS digital library does not provide only the dissertation document for each dissertation, but in some cases may also provide other additional documents, such as the examiners' reports etc. We evaluated our collaborative recommendation systems based on two different configurations of the user-by-item matrix: in the first configuration the value of each cell is non-binary – i.e. it contains the total number of downloads a user has performed for a particular dissertation; whilst in the second configuration the value of each cell is binary – 1 if the user has downloaded any document relating to the dissertations, and 0 otherwise.

The user-by-item ratings matrix is also decomposed using SVD to allow us to run the collaborative recommendation systems with LSA applied. The use of SVD renders the original user-by-item ratings matrix to be not sparse. The use of LSA is based on the premise that, since in content-based scenarios latent semantic features may be uncovered based on the co-occurrences of different terms, similar latent semantic features may also be uncovered on the basis of how different users 'like' similar items.

Two different algorithms were implemented that are able to perform collaborative recommendation. In the first algorithm, which we name *Collab-SimUsers*, those users that are similar to the user in question are identified. Then the set of candidate recommendations is constructed as the set of items that have not yet been rated by the current user, but that have been rated by similar users. The recommendation score of each candidate recommendations is the sum of the rating for that item by the similar user multiplied by the similarity of that user with the current user. This algorithm is given in pseudo-code below.

```
Collab-SimUsers (UserItemMatrix, CurUser)
{
    recommendations ← φ
    SimUsers ← findSimilarUsers(CurUser)
    CurUserRatings ← getUserRatings(CurUser)

    for each user in SimUsers
    {
        SimUserRatings ← getUserRatings(user)

        for each rating in SimUserRatings
        {
            if (rating ∉ CurUserRatings)
                recommendations ← recommendations ∪ rating

            Score(rating) ← Score(rating) +
                getUserSimilarity(CurUser, user)
        }
    }
}
```

The second collaborative recommendation algorithm (*Collab-SimItems*) works on the basis of item-to-item similarity, that is defined based on the overlap of the items' user vectors (each item is represented by the vector of users that have rated positively that item). A mean user vector is constructed from the user vectors of each item that has been rated positively by the selected user. This mean user vector is then compared to the user vectors of the other non-read items. Those items that have a sufficiently high similarity to the mean user vector are included in the list of recommendations. The list of recommendations is sorted on the basis of the similarity to the mean user vector. This listing below gives the pseudo-code for this algorithm.

```
Collab-SimItems (UserItemMatrix, CurUser)
{
    recommendations ← φ
    CurUserRatings ← getUserRatings(CurUser)
    CombVector⃗ ← φ

    for each ratedItem in CurUserRatings
    {
        RatingVect⃗ ← getUserRatingsVector(ratedItem)
        CombVector⃗ ← CombVector⃗ + RatingVect⃗
    }

    CombVector⃗ ← CombVector⃗ / |CurUserRatings|
```

```
for each item in Items
{
    if (item ∉ CurUserRatings)
    {
        itemVector ← getUserRatingsVector(item)
        Score(item) ← getSimilarity(CombVector, itemVector)

        if (Score(item) > SimThreshold)
            recommendations ← recommendations ∪ item
    }
}
}
```

LSA is applicable on both collaborative recommendation algorithms. In such cases, the decomposed user-by-item matrix is used instead of the original user-by-item ratings matrix.

Different similarity functions may be used to calculate the user similarity (in the case of *Collab-SimUsers* algorithm), or the item similarity (in the case of *Collab-SimItems* algorithms). These are Cosine Similarity, Pearson Similarity and Eucledian Distance.

4   Evaluation

As mentioned in Sect. 3.1, we used a set of download logs covering the period 20/2/2016 13:14:26 till 1/4/2016 08:50:53 to train the system, and a second set of download logs covering the period 2/4/2016 18:47:18 and 8/4/2016 10:49:59 to evaluate the system. The first set of downloads described 14443 download actions of 697 different dissertations by 5921 different users (not including search-engine bots). The evaluation set of downloads details 3145 downloads actions of 478 dissertations by 2152 different users.

The evaluation metric used was recall – i.e. the fraction of downloads in the evaluation set (of previously undownloaded dissertations by users that also performed some download in the training set) that have also been recommended by our systems. One has to note that this evaluation is not optimal, especially when considering that a user may not be aware of 'interesting' dissertations that are recommended by our systems. However, we considered this evaluation to be a suitable indicator of how the different algorithms compare to each other.

For this evaluation, we had each of our systems output the top 20 recommendations, and then calculated recall on the basis of these 20 recommendations. As a baseline, we had a system that generated a number of random recommendations for each user for previously undownloaded dissertations.

We evaluated the 27 recommendation configurations listed below, and results are shown in Table 1 below:

- Baseline random recommendation
- Content-based recommendation

  - With and Without LSA

- *Collab-SimUsers* Collaborative Recommendation:

  - With and Without LSA
  - Using binary ratings, and non-binary ratings
  - Using Cosine, Pearson and Eucledian Similarity

Table 1. Evaluation results

System Description	LSA	Recall
Random Recommendation		0.051
Content-based Recommendation	No LSA	0.200
Content-based Recommendation	k = 150	0.232
Collab-SimUsers, non-binary ratings, Cosine Similarity	No LSA	0.291
Collab-SimUsers, non-binary ratings, Cosine Similarity	k = 50	0.369
Collab-SimUsers, binary ratings, Cosine Similarity	No LSA	0.261
Collab-SimUsers, binary ratings, Cosine Similarity	k = 50	0.41
Collab-SimUsers, non-binary ratings, Pearson Similarity	No LSA	0.291
Collab-SimUsers, non-binary ratings, Pearson Similarity	k = 50	0.338
Collab-SimUsers, binary ratings, Pearson Similarity	No LSA	0.272
Collab-SimUsers, binary ratings, Pearson Similarity	k = 50	**0.416**
Collab-SimUsers, non-binary ratings, Eucledian Similarity	No LSA	0.063
Collab-SimUsers, non-binary ratings, Eucledian Similarity	k = 50	0.185
Collab-SimUsers, binary ratings, Eucledian Similarity	No LSA	0.072
Collab-SimUsers, binary ratings, Eucledian Similarity	k = 50	0.338
Collab-SimItems, non-binary ratings, Cosine Similarity	No LSA	0.382
Collab-SimItems, non-binary ratings, Cosine Similarity	k = 50	0.311
Collab-SimItems, binary ratings, Cosine Similarity	No LSA	0.355
Collab-SimItems, binary ratings, Cosine Similarity	k = 50	0.294
Collab-SimItems, non-binary ratings, Pearson Similarity	No LSA	0.386
Collab-SimItems, non-binary ratings, Pearson Similarity	k = 50	0.294
Collab-SimItems, binary ratings, Pearson Similarity	No LSA	0.358
Collab-SimItems, binary ratings, Pearson Similarity	k = 50	0.299
Collab-SimItems, non-binary ratings, Eucledian Similarity	No LSA	0.032
Collab-SimItems, non-binary ratings, Eucledian Similarity	k = 50	0.076
Collab-SimItems, binary ratings, Eucledian Similarity	No LSA	0.023
Collab-SimItems, binary ratings, Eucledian Similarity	k = 50	0.072

– *Collab-SimItems* Collaborative Recommendation:

- With and Without LSA
- Using binary ratings, and non-binary ratings
- Using Cosine, Pearson and Eucledian Similarity

From the results, one can note that on the whole, collaborative based recommendation generally performs better than content-based recommendation (as reported also by [10]). The best performing system is *Collab-SimUsers* that works on binary ratings using LSA and Pearson similarity, with a recall of approximately 42 %. One can also note that the best performing similarity function is Pearson similarity with Cosine similarity scoring a close second. Eucledian distance seems to be particularly unsuitable for these collaborative algorithms. On the whole, using binary ratings produces better results than non-binary ratings.

One can note also the positive effect of LSA. LSA improves scores in content-based recommendation, and also in the *Collab-SimUsers* collaborative setup. On the other hand, it degrades the scores in the *Collab-SimItems* (with some minor exceptions). The best number of dimensions used (k) is consistently at 50 when using collaborative recommendation techniques. These results indicate that LSA can be used as a possible solution to the sparse matrix problem using the top 50 dimensions.

Apart from considering the considerable difference of results between content-based and collaborative recommendation systems, one should also consider the fact that the collaborative recommendation systems operate much faster than the content-based recommendation system used. This further solidifies the argument that collaborative techniques are more suitable for calculating recommendations within the PHD UNS digital library.

5   Conclusions and Future Work

In this research, we present and compare a number of different algorithms that perform personalised recommendations for users of the PHD UNS digital library. Apart from comparing content-based and collaborative approaches, we also evaluate the use of LSA and the use of different similarity functions. Results show that the most suitable approach for this task is the collaborative approach that works on the basis of user similarity, applies LSA and utilises Pearson similarity.

One should note however, that the evaluation described here is not exhaustive enough. It does not provide any indication on the accuracy of relevant recommendations. A user may not be aware at all of relevant dissertations. Therefore, the fact that a user has not downloaded a dissertation does not mean that he/she is not interested in it. A better evaluation can be performed by having the system generate recommendations for a user, and have the user provide feedback as to whether he/she are interested in the recommended dissertations. Otherwise, this may be performed more implicitly by logging the extent to which users are utilising the generated lists of recommendations.

Future research can involve the implementation and evaluation of a hybrid system – whereby the content-based system is combined with the best performing collaborative approach. The best performing system will be incorporated within the PHD UNS digital library. Further plans also include the utilisation of recommendation algorithms to re-rank search results thus providing personalised search.

Acknowledgements. This research was initiated through a Short Term Scientific Mission (STSM) financed by the IC1302 KEYSTONE (semantic KEYword-based Search on sTructured data sOurcEs) COST Action (http://www.keystone-cost.eu/.).

References

1. Albayrak, S., Wollny, S., Varone, N., Lommatzsch, A., Milosevic, D.: Agent technology for personalized information filtering: the PIA system. In: SAC 2005: Proceedings of the 2005 ACM Symposium on Applied Computing, pp. 54–59. ACM, New York (2005)
2. Azzopardi, J., Staff, C.: Automatic adaptation and recommendation of news reports using surface-based methods. In: Pérez, J., et al. (eds.) PAAMS. AISC, vol. 156, pp. 69–76. Springer, Heidelberg (2012)
3. Belkin, N.J., Croft, W.B.: Information filtering and information retrieval: two sides of the same coin? Commun. ACM **35**(12), 29–38 (1992)
4. Bordogna, G., Pagani, M., Pasi, G., Villa, R.: A flexible news filtering model exploiting a hierarchical fuzzy categorization. In: Larsen, H.L., Pasi, G., Ortiz-Arroyo, D., Andreasen, T., Christiansen, H. (eds.) FQAS 2006. LNCS (LNAI), vol. 4027, pp. 170–184. Springer, Heidelberg (2006). doi:10.1007/11766254_15
5. Callan, J.: Learning while filtering documents. In: SIGIR 1998, Proceedings of the 21st Annual International ACM SIGIR Conference on Research and Development in Information Retrieval, pp. 224–231. ACM, New York (1998)
6. Choi, Y.S.: Content type based adaptation in collaborative recommendation. In: Proceedings of the 2014 Conference on Research in Adaptive and Convergent Systems, RACS 2014, pp. 61–65. ACM, New York (2014)
7. Das, A.S., Datar, M., Garg, A., Rajaram, S.: Google news personalization: scalable online collaborative filtering. In: WWW 2007, Proceedings of the 16th International Conference on World Wide Web, pp. 271–280. ACM, New York (2007)
8. Deerwester, S., Dumais, S.T., Furnas, G.W., Landauer, T.K., Harshman, R.: Indexing by latent semantic analysis. J. Am. Soc. Inform. Sci. **41**(6), 391–407 (1990)
9. Foltz, P.W., Dumais, S.T.: Personalized information delivery: an analysis of information filtering methods. Commun. ACM **35**(12), 51–60 (1992)
10. Garcin, F., Zhou, K., Faltings, B., Schickel, V.: Personalized news recommendation based on collaborative filtering. In: Proceedings of the The 2012 IEEE/WIC/ACM International Joint Conferences on Web Intelligence and Intelligent Agent Technology, WI-IAT 2012, vol. 01, pp. 437–441. IEEE Computer Society, Washington, DC (2012)
11. Lang, K.: Newsweeder: learning to filter netnews. In: ML 1995, Proceedings of the 12th International Machine Learning Conference, pp. 331–339. Morgan Kaufman (1995)

12. Ling, G., Lyu, M.R., King, I.: Ratings meet reviews, a combined approach to recommend. In: Proceedings of the 8th ACM Conference on Recommender Systems, RecSys 2014, pp. 105–112. ACM, New York (2014)
13. Middleton, S.E., De Roure, D.C., Shadbolt, N.R.: Capturing knowledge of user preferences: ontologies in recommender systems. In: Proceedings of the 1st International Conference on Knowledge Capture, K-CAP 2001, pp. 100–107. ACM, New York (2001)
14. Morita, M., Shinoda, Y.: Information filtering based on user behavior analysis and best match text retrieval. In: Croft, B.W., van Rijsbergen, C.J. (eds.) SIGIR 1994, pp. 272–281. Springer, New York (1994)
15. Pasi, G., Bordogna, G., Villa, R.: A multi-criteria content-based filtering system. In: SIGIR 2007: Proceedings of the 30th Annual International ACM SIGIR Conference on Research and Development in Information Retrieval, pp. 775–776. ACM, New York (2007)
16. Pazzani, M.J., Billsus, D.: Content-based recommendation systems. In: Brusilovsky, P., Kobsa, A., Nejdl, W. (eds.) The Adaptive Web. LNCS, vol. 4321, pp. 325–341. Springer, Heidelberg (2007). doi:10.1007/978-3-540-72079-9_10
17. Pon, R.K., Cardenas, A.F., Buttler, D., Critchlow, T.: iScore: measuring the interestingness of articles in a limited user environment. In: CIDM 2007: Proceedings of the IEEE Symposium on Computational Intelligence and Data Mining, pp. 354–361 (2007)
18. Pon, R.K., Cardenas, A.F., Buttler, D., Critchlow, T.: Tracking multiple topics for finding interesting articles. In: KDD 2007, Proceedings of the 13th ACM SIGKDD International Conference on Knowledge Discovery and Data Mining, pp. 560–569. ACM, New York (2007)
19. Pon, R.K., Cárdenas, A.F., Buttler, D.J.: Online selection of parameters in the rocchio algorithm for identifying interesting news articles. In: WIDM 2008, Proceedings of the 10th ACM Workshop on Web Information and Data Management, pp. 141–148. ACM, New York (2008)
20. Schafer, J.B., Frankowski, D., Herlocker, J., Sen, S.: Collaborative filtering recommender systems. In: Brusilovsky, P., Kobsa, A., Nejdl, W. (eds.) The Adaptive Web. LNCS, vol. 4321, pp. 291–324. Springer, Heidelberg (2007). doi:10.1007/978-3-540-72079-9_9

Keyword-Based Search
on Bilingual Digital Libraries

Ranka Stanković[1(✉)], Cvetana Krstev[2], Duško Vitas[3], Nikola Vulović[1],
and Olivera Kitanović[1]

[1] Faculty of Mining and Geology, University of Belgrade, Belgrade, Serbia
{ranka,ranka.stankovic,nikola.vulovic,olivera.kitanovic}@rgf.bg.ac.rs
[2] Faculty of Philology, University of Belgrade, Belgrade, Serbia
cvetana@matf.bg.ac.rs
[3] Faculty of Mathematics, University of Belgrade, Belgrade, Serbia
vitas@matf.bg.ac.rs

Abstract. This paper outlines the main features of Bibliša, a tool that
offers various possibilities of enhancing queries submitted to large collections of aligned parallel text residing in bilingual digital library. Bibliša
supports keyword queries as an intuitive way of specifying information needs. The keyword queries initiated, in Serbian or English, can
be expanded, both semantically, morphologically and in other language,
using different supporting monolingual and bilingual resources. Terminological and lexical resources are of various types, such as wordnets, electronic dictionaries, SQL and NoSQL databases, which are distributed in
different servers accessed in various ways. The web application has been
tested on a collection of texts from 3 journals and 2 projects, comprising
299 documents generated from TMX, stored in a NoSQL database. The
tool allows the full-text and metadata search, with extraction of concordance sentence pairs for translation and terminology work support.

1   Introduction

In this paper, we outline the main features of Bibliša[1], a tool developed within
the Human Language Technology group at the University of Belgrade, aimed at
enhancement of search possibilities in bilingual digital libraries (DL). The tool
offers cross-lingual search functions for large collections of aligned texts, which
enable users to compose queries both with simple and multiword keywords in
more than one language. In addition to that, the system can expand user queries
both semantically and morphologically, the latter being very important in highly
inflective languages, such as Serbian.

This work extends our previous research [12] where solution was developed for
only one parallelized journal with 7,986 segments and queries could be expanded
using tree multilingual resources. Apart from enlarging the existing text collection that now comprises 13,475 segments, we have introduced 5 new text collections with 71,310 segments and enhanced substantially web application with

---

[1] http://hlt.rgf.bg.ac.rs/Biblisha.

© Springer International Publishing AG 2017
A. Calì et al. (Eds.): IKC 2016, LNCS 10151, pp. 112–123, 2017.
DOI: 10.1007/978-3-319-53640-8_10

metadata editor, statistical module, lexical analysis, bag of words and tag cloud for each document, new filtering option of concordances, etc. and supporting database organization. The number of used lexical resources is also increased by 3 new term databases and the quality of previously used resources is improved.

Our focus is on bilingual scientific open-access journals with articles both in Serbian and in English. The first text collection was INFOtheca journal, which covers the field of Digital Humanities. Later three more journals and resources produced within two project, all with text published both in Serbian and in English, were added.

The first implementation of enhanced cross-lingual search, HLT group implemented only for one text and later for one text collection. The tool for lexical resource management LeXimir was developed within the group, with the aim to build an integrated environment for development and maintenance of various resources, such as morphological e-dictionaries and wordnets. Along with various dictionary management modules, LeXimir has modules for multiword unit extraction and handling of both monolingual and aligned texts. Enhanced querying of aligned texts is based on available lexical resources to perform semantic and morphological expansion of queries. The tool was, however, unsuitable for large collections of documents such as bilingual digital libraries of e-journals. In order to enable cross-lingual search to large collections we had to turn to NoSQL databases, which handles storage and indexing of collections of XML or JSON documents. Thus, the initial version of the new tool that extends LeXimir's capabilities to XML databases, named Bibliša, was developed [12].

Section 2 gives a brief overview of the related work on bilingual digital libraries and keyword-based search on different structures and types of texts. In Sect. 3 we present the structure and content of text collections in Bibliša's digital library, while the overview of lexical resources used as a support for linguistic analysis and query expansion is given in Sect. 4. The software solution details are presented in Sect. 5 with a stress on significant improvements that have been introduced over last few years. Section 6 concludes the paper and outlines the future work.

2   Related Work

In this section, we give a brief overview of previous work related to the search of bilingual parallel documents and the analysis of the query expansion effects. Results of these studies show that the use of lexical resources can substantially improve the retrieval recall of keyword-based search on bilingual digital libraries as well as the formulation of queries.

Gravano and Henzinger patented the system and methods that use parallel corpora to translate terms from a search query in one language to another language by using the content of retrieved and linked documents [3]. OPUS, as a language resource of parallel corpora and related tools, provides freely available data sets in various formats together with basic annotation to be used for applications in computational linguistics, translation studies and cross-linguistic corpus studies [16].

Thong et al. gave various recommendations for increasing user acceptance of digital libraries. They recognise three categories of external factors: interface characteristics, organizational context and the user need difference. Interface characteristics comprise terminology clarity, screen design, and navigation clarity, while organizational context includes relevance, system accessibility, and system visibility. User-centered design tries to optimize use of digital libraries around how users can, want, or need to use it. A good screen design can create a comfortable environment where users can easily identify functional groups and navigation aids, move around and scan search results, and make more efficient searches [15].

Knowledge is recognized as the ultimate resource of our days while digital libraries provide extraordinary information and knowledge highways for all. Authors in [6] outline the need for building library systems that go a step beyond the existing technological infrastructure and present an attempt to clarify the main directions for research on the concept of a "semantic digital library", and the main management and technical challenges derived from such idea.

Volk [18] argue that automatic word alignment allows for major innovations in searching parallel corpora. The Multilingwis (Multilingual Word Information System) contains texts in five languages from Europarl3 with cross-language alignments down to the word level and allows a user to search for single words or multi-word expressions and returns the corresponding translation variants in the four other languages [2].

3   The Digital Library

Bibliša's digital library consists of six text collections, four of them being scientific journals and two resources produced within international projects. Text documents from our digital library fall in the category of unstructured and semi-structured collections of data. Usually, a collection of such documents is supported by metadata content, meaning that the data manipulation approach has to be different from the approach in classical relational databases.

Bibliša is aimed at search of document collections consisting of aligned parallel texts converted in TMX (Translation Memory eXchange) format. TMX is an open XML-based standard intended for easier exchange of translation memory data, that is, aligned parallel texts, between tools and translation vendors [9]. A TMX document consists of a header and a body. The header consists of metadata describing the aligned texts, while the body contains a set of translation units (TU) with two or more translation unit variants (TUV), containing a same text segment in different languages. In our text collection, TUVs segments correspond to sentences, but a segment can also be a part of a sentence or can constitute of several sentences. Each TUV has an attribute xml:lang with ISO code of the language of the segment text as its value.

The first TMX document collection implemented in Bibliša was generated from journal articles in Digital Humanities domain published in INFOtheca[2].

---

[2] http://infoteka.bg.ac.rs/index.php/en/.

Next, three more journals were added: Underground Mining Engineering[3] and Journal of architecture, urbanism and spatial planning[4]. With TEMPUS project Blending Academic and Entrepreneurial Knowledge in Technology Enhanced Learning – BAEKTEL[5] we started with parallelisation of deliverable reports that have been published bilingually [10]. A result from INTERA project (Integrated European language data Repository Area), a collection of aligned texts from education, finance, health and law domain [1] were recently added. The addition of articles from some other journals are in plan. The overview of statistics on text collections presented in Bibliša is given in Table 1.

Table 1. The overview of the text collection statistics

Collection	Number of documents	Number of sentences
INFOtheca	74	13, 457
Underground Mining Engineering	55	4, 381
Architecture and urbanism	10	1, 641
BAEKTEL	3	915
INTERA	157	46, 497
Management	110	17, 426
Total	409	84, 767

For TMX document production the development environment ACIDE [7] is used, which incorporates Xalign tool for automatic sentence alignment[6] and Concordancier for alignment visualization and manual correction of alignment errors, both developed by LORIA (Laboratoire lorrain de recherche en informatique et ses applications). The first language of an aligned text in our digital library is always English and the second language is Serbian, regardless of what the source language (being translated from) and target languages (translated to) are. The example in Fig. 1 shows a translation unit from an Underground Mining Journal article: a translation unit variant in English and its corresponding TUV in Serbian in two different formats: XML and JSON. Original TMX format is in XML, but our system supports its automatic transformation into JSON without information loss.

Bibliša's digital library contains metadata in XML and JSON format (Fig. 2), with details about articles and journals, or deliverables and projects. It comprises metadata grouped in Collection, Journal and Document sections. Structure of content data is grouped in TMX collection, while TmxBow represents Bag of Words as document vector used for document ranking [8,11].

---

[3] http://www.rgf.bg.ac.rs/publikacije/PodzemniRadovi.
[4] http://www.iaus.ac.rs/code/navigate.aspx?Id=221.
[5] http://www.baektel.eu/.
[6] http://led.loria.fr/outils/ALIGN/align.html.

```
<tu>
  <prop type="Domain">Tomašević et al., 2012, vol. XX:20, ID: 2.2012.20.4
  </prop>
  <tuv xml:lang="en" creationid="n5 " creationdate="20140525T090842Z">
    <seg>It is the authors' wish to bring closer to the mining public at
    least a small part of the possibilities offered by GIS, both in mining
    activities and in environment protection. </seg>
  </tuv>
  <tuv xml:lang="sr" creationid="n5 " creationdate="20140525T090842Z">
    <seg>Želja autora je da rudarskoj javnosti približi samo mali deo
    ogromnih mogućnosti koje GIS pruža, kako pri rudarskim aktivnostima,
    tako i u zaštiti životne sredine. </seg>
  </tuv>
</tu>
```
```
"Desc": "Tomašević et al., 2012, vol. XX:20, ID: 2.2012.20.4",
"SentenceID": "n5",
"tuv": [{
    "lang": "en",
    "seg": "It is the authors' wish to bring closer to the mining
    public at least a small part of the possibilities offered by
    GIS, both in mining activities and in environment protection. "
},
{
    "lang": "none",
    "seg": "Želja autora je da rudarskoj javnosti približi samo mali
    deo ogromnih mogućnosti koje GIS pruža, kako pri rudarskim
    aktivnostima, tako i u zaštiti životne sredine. "
}]
```

Fig. 1. A TU with a segment in English and its corresponding segment in Serbian in two different formats.

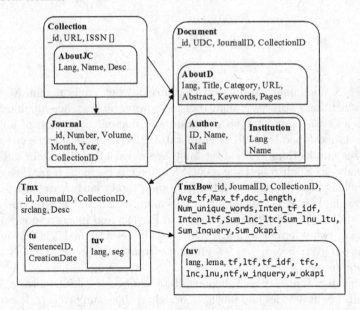

Fig. 2. An overview of the content and metadata structure

All metadata, except language independent data, such as numerical metadata (ID, Number, Volume, Month, Year), the UDC and Mail, are entered in both languages (Serbian and English), using the attribute @xml:lang for XML serialization, or lang for JSON serialization, to denote the language of the content.

These metadata can be used both for the refinement of full text document search and for additional forms of searching and browsing, as illustrated in Sect. 5.

4 Supporting Resources

The overall requirements for the semantic approach to digital libraries are the use of ontologies, lexical and terminological resources. The development of a digital library tool for search and browse, based on the reports and recommendations in literature and our previous experience.

Three types of lexical resources are used for the expansion of queries submitted to our collection of documents. The most important resources are Serbian morphological electronic dictionaries of simple words and multi-word units [5]. The dictionaries are used to generate all inflective forms of query keywords, thus improving the system recall without negative effects on precision. Moreover, for multi-word keywords not found in dictionaries, there exists a rule-based strategy, which attempts to recognize their syntactic structure and produce inflected forms [13]. The lexical database is used for the fast query expansion with inflected forms.

Another type of resources used by Bibliša for bilingual and semantic expansion of queries are wordnets. At present we use the Princeton English Wordnet (PWN), version 3.0, as well as the Serbian Wordnet (SrpWN), initially developed in the scope of the BalkaNet project [17] and subsequently enhanced and upgraded. Most of the synsets in Serbian Wordnet are aligned with PWN synsets via the Interlingual Index, with the exception of Serbian specific synsets that are not available in PWN. The Interlingual Index enables bilingual query expansion, while synonymous relations expressed by synsets themselves enable semantic expansion. At present, we do not use all semantic relations available in both wordnets, just synonyms, hyponyms and hypernyms.

Bibliša also uses various domain specific bilingual dictionaries and termbases. One of dictionaries used is "Dictionary of librarianship: English-Serbian and Serbian-English" [4]. GeolISSTerm and RudOnto were developed at the Faculty of Mining and Geology, University of Belgrade [14]. GeolISSTerm is a thesaurus of geological terms with entries in Serbian and English. RudOnto is another complex terminological resource, being developed at the same faculty with the aim of becoming the future e-format reference resource in Serbian for mining terminology. Currently RudOnto comprises terms in Serbian, their English equivalents, and a small number of equivalents in other languages. The Termi[7] application has recently been launched to serve as a support for the development of terminological dictionaries in various domains. All mentioned resources are used to improve the full-text search performance of our DL, especially the recall.

5 Software Implementation

Development of NoSQL (Not only SQL) databases was a response to immense growth of data in unstructured and semi-structured format. NoSQL encompasses

---

[7] http://termi.rgf.bg.ac.rs.

a wide variety of different database technologies that were developed for management of massive volumes of new, rapidly changing data types – structured, semi-structured, unstructured and polymorphic data. The NoSQL are not expected to substitute relational databases, but rather to offer a tool for development of more-robust systems that enable efficient management of loosely structured documents.

In the initial face of our research, we used eXist-db, but later we included also MarkLogic,[8] and we found the latter as more robust and stable platform. Recently, we started using MongoDB, as a document-based data model with the basic unit of storage analogous to JSON. MongoDB supports dynamic queries on documents using a powerful document-based query language. It has huge user community and support from large companies that assures sustainability. Since our university, as many other universities, cannot support commercial licenses for research, we are now turning to scale-out architectures of our software solutions using open source software as platform for development.

The Bibliša tool operates within a complex system composed of several modules as depicted in Fig. 3. It is oriented towards textual collections of aligned documents and the corresponding metadata described in Sect. 3, whereby it uses several other lexical resources described in Sect. 4.

Bibliša's user formulates the initial query in the form of one or more simple or multiword keywords, which are then forwarded to the web service Vebran for

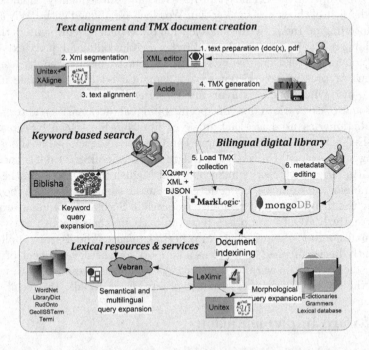

Fig. 3. Bibliša+LeXimir component model

---

[8] http://www.marklogic.com.

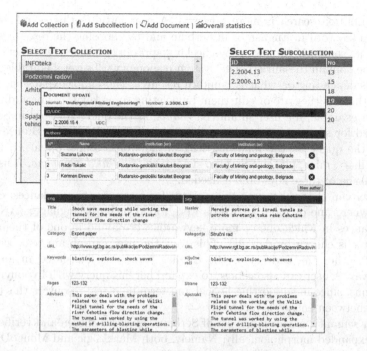

Fig. 4. The interface for metadata editing

further semantic expansion, depending on the user's preferences [13]. The web service, which receives the query from the web application, invokes the LeXimir function library and Unitex[9] routines to perform query expansion according to user specifications, and returns the result to the web service, which in its turn returns it to the web application that invoked him. When the expanded query reaches Bibliša, the tool forwards it to the document collection in the form of an XQuery. The results, a set of aligned concordances, are formatted and presented to the user. The concordances are preceded by information identifying the document they originate from, and a link to summary metadata for this document in both languages.

A user can search one or several collections in two different ways, using metadata or full-text search. When performing monolingual search of metadata, a user can make use of a form with predefined fields for the most commonly used data: author's name, words from an article title, year of publication, and article keywords. As a result, a list of all articles in the language of the query that fulfill the required conditions is produced with all available metadata. In addition to that, for each article, links are offered to the full text of the article in .pdf format as well as an excerpt of the aligned text of the article in .html format. The new component for metadata management has been developed in order to enable multiuser and distributed metadata editing from different institutions (Fig. 4).

---

[9] The corpus processing system http://igm.univ-mlv.fr/~unitex.

The full-text search is more powerful. A user initiates this search by submitting a keyword in English or in Serbian and by choosing the resources to be used for its semantic expansion by checking appropriate boxes (wordnet and/or one of the domain thesaurus). The system responds with several editable lists of keywords depending on what is found in resources chosen for expansion. A user can use them as they are or edit them by deleting some keywords or adding new ones. For example, if a user submitted *rudnik*, as the query keyword in Serbian, and asked for semantic expansion using wordnets and RudOnto, Bibliša would expand the query by the following English keywords: mine, open pit, surface mine, colliery, mine, pit, and Serbian synonyms *površinski kop, okno*. A user can remove some of these keywords if needed.

A user can choose between two options for retrieving concordances: concordances where one or more keywords were found in both languages (AND), or concordances in which one or more keywords were found in one of them (OR) (the latter is chosen by default). This way, the user can obtain text segments in which either the Serbian or the English term was translated in an unexpected way. A user can also choose to expand her/his query with hyponyms and hypernyms, since in supporting lexical and terminological resources this type of relation between entries is recorded.

After semantic expansion, a set of Serbian keywords chosen and verified by a user is expanded morphologically. Namely, both MarkLogic and MongoDB support stemming for several languages including English; however, Serbian is not among them. Thus, the morphological forms of English keywords are taken care of by MarkLogic's and MongoDB's stemming capability. As for Serbian, Bibliša must expand the initial query with all morphological forms of the keywords using the available Serbian resources (see Sect. 4) [5].

The semantically and morphologically expanded query is then used to search through TUVs taking into account the value of the xml:lang attribute or lang property. Namely, English keywords will search only through TUVs with xml:lang or lang attribute set to "en", while Serbian keywords will search through TUVs in MarkLogic database with xml:lang attribute set to "sr", or in MongoDB set to "none", since MongoDB does not support text indexing if language code is set to "sr".

To illustrate morphological expansion of a query in Bibliša we will use the keyword *površinski kop* (open pit). Morphological expansion is performed automatically and no intervention form a user is expected. The expanded query consists of several Serbian entries for one term (several inflected forms), but only one for English entry. Part of the XQuery code that follows is used for transforming the expanded query to the proper XQuery form:

```
cts:or-query((cts:word-query("povrsinski kop",("stemmed","lang=sr")),
cts:or-query((cts:word-query("povrsinskog kopa",("stemmed","lang=sr")),
cts:or-query((cts:word-query("povrsinskom kopu",("stemmed","lang=sr")),
cts:or-query((cts:word-query("povrsinski kopovi",("stemmed","lang=sr")),
.....
cts:word-query("open pit",("stemmed","lang=en")
```

Starting from the initial query, processed as outlined, the system finally generates aligned concordances in which all retrieved keywords are highlighted in both languages. Each concordance line is preceded by an identification of the document it originates from which contains a link to the full text of a document form which it was extracted. Within this identification is a link to the full metadata of a retrieved document. A few concordance lines for the initial query *rudnik* 'mine' are presented in Fig. 5.

Miladinović et al., 2011, No. 19, ID: 2.2011.19.4 metadata	n31 : There are currently many professional software packages, which include economic evaluation of open pit *mine*, the geology of the ore body, transport communications and other technological processes.	n31 : Danas već postoji mnoštvo profesionalnih programskih paketa koji obuhvataju ekonomsku ocenu *površinskog kopa*, geologiju rezervi, transportne komunikacije i sve druge tehnološke procese.
Miladinović et al., 2011, No. 19, ID: 2.2011.19.4 metadata	n33 : There is almost no *surface mine* where engineers do not use any of the computer programs to assist in the implementation of specific project solutions.	n33 : Gotovo da ne postoji *površinski kop* gde inženjeri ne koriste neki od računarskih programa kao pomoć pri izvođenju određenih projektnih rešenja.
Miladinović et al., 2011, No. 19, ID: 2.2011.19.4 metadata	n42 : Software package Gemcom is designed for geological interpretation and modeling of unstratified deposits and the design of surface and underground *mining* of metals and nonmetals.	n42 : Programi Gemcom se primenjuju za geološku interpretaciju i modeliranje neslojevitih ležišta i projektovanje površinskih i podzemnih *rudnika* metala i nemetala.
Miladinović et al., 2011, No. 19, ID: 2.2011.19.4 metadata	n65 : This programme represent a standard for optimization of open-pit *mines*, or harmonizing of financial viability and optimal exploitation strategy for the open-pit.	n65 : Predstavlja standard za optimizaciju površinskih kopova, odnosno za usklađivanje finansijske isplativosti i optimalne strategije eksploatacije na *površinskom kopu*.

Fig. 5. Concordances for the initial query *rudnik*

An example of expanding the initial query *povreda* 'injury' is presented in Fig. 6 (top), with number of retrieved segments for the initial query itself, followed by several types of expansion: morphological, semantic, morphological and semantic, semantic and English and all tree types of expansion (full). Retrieval with full expansion is assumed to have recall 1 and other results are calculated as relative to it. Manual evaluation was performed for 24 queries, half of them

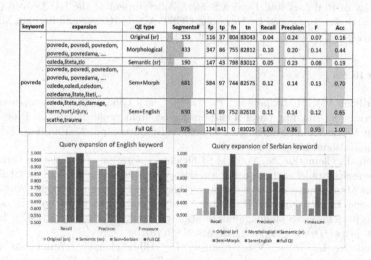

Fig. 6. Evaluation of query expansion

with Serbian and half with English initial keywords, and recall, precision and F-measure were calculated for all of them. Evaluation of query expansion benefits were analyzed and average performance is presented in Fig. 5 (bottom). Results show that morphological expansion increases the response, while maintaining precision. With semantic expansion which increases the number of results, the precision can drop. Query expansion to another language improves results on the average.

6   Future Work

The performance of Bibliša, our tool enabling enhanced search of bilingual digital libraries of text collections, has been tested on a TMX document collection generated from 409 documents and their translations published in four Serbian journals and produced by two projects. We hope that the improved web application for metadata management together with new search capabilities will enable the enlargement of our collection and lead to the greater number of users. We plan to expand the set of used lexical resources with other multilingual resources and to exploit various semantic relations offered by them in order to achieve a more versatile semantic expansion. For example, the use of derivation relation would enable us to expand a query *rudnik* (a noun 'mine') with *rudarski* (an adjective 'mining') thus improving the system's recall. We began the first experiments with the word alignment and we plan to include the obtained lists into the query expansion support. Another ongoing activity is the improvement of the search result visualization including statistics on documents in the form of interactive frequency histograms, tag clouds and document clustering.

Acknowledgements. Preprocessing of texts and correction of the alignment were done by Biljana Lazić, Jelena Andonovski and Jelena Andjelković, PhD students at the Faculty of Philology and Danica Seničić, MSc student at the KU Leuven (LLN). This research was supported by Keystone COST Action IC1302 and Serbian Ministry of Education and Science under the grant #III 47003.

References

1. Gavrilidou, M., Labropoulou, P., Desipri, E., Giouli, V., Antonopoulos, V., Piperidis, S.: Building parallel corpora for econtent professionals. In: Proceedings of the Workshop on Multilingual Linguistic Resources, pp. 97–100. Association for Computational Linguistics (2004)
2. Graën, J., Clematide, S., Volk, M.: Efficient exploration of translation variants in large multiparallel corpora using a relational database. In: 4th WS on Challenges in the Management of Large Corpora (Workshop Programme), p. 20 (2016)
3. Gravano, L., Henzinger, M.H.: Systems and methods for using anchor text as parallel corpora for cross-language information retrieval, US Patent 8,631,010, January 2014. https://www.google.ch/patents/US8631010
4. Kovačević, L., Injac, V., Begenišić, D.: Bibliotekarski terminološki rečnik: englesko-srpski, srpsko-engleski [Library Terminological Dictionary: English-Serbian, Serbian-English]. Narodna biblioteka Srbije (2004)

5. Krstev, C.: Processing of Serbian – Automata, Texts and Electronic Dictionaries. Faculty of Philology, University of Belgrade, Belgrade (2008)
6. Lytras, M., Sicilia, M.A., Davies, J., Kashyap, V., Lytras, M., Sicilia, M.A., Davies, J., Kashyap, V.: Digital libraries in the knowledge era: knowledge management and semantic web technologies. Libr. Manage. **26**(4/5), 170–175 (2005)
7. Obradović, I., Stanković, R., Utvić, M.: An integrated environment for development of parallel corpora. In: Die Unterschiede zwischen dem Bosnischen/Bosniakischen, Kroatischen und Serbischen, pp. 563–578 (2008)
8. Radovanović, M., Ivanović, M.: Text mining: approaches and applications. Novi Sad J. Math. **38**(3), 227–234 (2008)
9. Savourel, Y.: TMX 1.4 b Specification, The Localisation Industry Standards Association (LISA) (2004)
10. Stanković, R., Krstev, C., Lazić, B., Vorkapić, D.: A bilingual digital library for academic and entrepreneurial knowledge management. In: Spender, J., Schiuma, G., Albino, V. (eds.) 10th International Forum on Knowledge Asset Dynamics – IFKAD 2015, pp. 1764–1777 (2015). http://www.knowledgeasset.org/Proceedings/
11. Stanković, R., Krstev, C., Obradović, I., Kitanović, O.: Indexing of textual databases based on lexical resources: a case study for Serbian. In: Cardoso, J., Guerra, F., Houben, G.-J., Pinto, A.M., Velegrakis, Y. (eds.) KEYSTONE 2015. LNCS, vol. 9398, pp. 167–181. Springer, Heidelberg (2015). doi:10.1007/978-3-319-27932-9_15
12. Stanković, R., Krstev, C., Obradović, I., Trtovac, A., Utvić, M.: A tool for enhanced search of multilingual digital libraries of e-journals. In: Proceedings of the 8th International Conference on Language Resources and Evaluation (LREC 2012) (2012)
13. Stanković, R., Obradović, I., Krstev, C., Vitas, D.: Production of morphological dictionaries of multi-word units using a multipurpose tool. In: Jassem, K., Fuglewicz, P.W., Piasecki, M., Przepiórkowski, A. (eds.) Proceedings of the Computational Linguistics-Applications Conference, pp. 77–84 (2011). ISBN: 978-83-60810-47-7
14. Stanković, R., Trivić, B., Kitanović, O., Blagojević, B., Nikolić, V.: The Development of the GeolISSTerm Terminological Dictionary. INFOtheca **12**(1), 49a–63a (2011)
15. Thong, J.Y., Hong, W., Tam, K.Y.: What leads to user acceptance of digital libraries? Commun. ACM **47**(11), 78–83 (2004)
16. Tiedemann, J.: Parallel data, tools and interfaces in OPUS. In: Proceedings of the 8th International Conference on Language Resources and Evaluation (LREC 2012) (2012)
17. Tufis, D., Cristea, D., Stamou, S.: Balkanet: aims, methods, results and perspectives. A general overview. Rom. J. Inf. Sci. Technol. **7**(1–2), 9–43 (2004)
18. Volk, M., Graën, J., Callegaro, E.: Innovations in parallel corpus search tools. In: Proceedings of the 9th International Conference on Language Resources and Evaluation (LREC 2014), pp. 3172–3178 (2014)

Network-Enabled Keyword Extraction
for Under-Resourced Languages

Slobodan Beliga[(⊠)] and Sanda Martinčić-Ipšić

Department of Informatics, University of Rijeka,
Radmile Matejčić 2, 51 000 Rijeka, Croatia
{sbeliga, smarti}@inf.uniri.hr

Abstract. In this paper we discuss advantages of network-enabled keyword extraction from texts in under-resourced languages. Network-enabled methods are shortly introduced, while focus of the paper is placed on discussion of difficulties that methods must overcome when dealing with content in under-resourced languages (mainly exhibit as a lack of natural language processing resources: corpora and tools). Additionally, the paper discusses how to circumvent the lack of NLP tools with network-enabled method such is SBKE method.

Keywords: Network-enabled keyword extraction · Under-resourced languages · NLP tools · SBKE method

1 Introduction

Automatic keyword extraction is the process of identifying key terms, phrases, segments or words from a textual content that can appropriately represent the main topic of the document [1, 14]. Keyword extraction (KE) methods can be roughly divided into three categories: supervised, semi-supervised and unsupervised [1]. Network-enabled or graph-based are considered as unsupervised KE methods.

Today the automatic keyword extraction from texts still remains an open question, especially for content written in under-resourced languages. For under-resourced languages there are no reliable tools which can be used for keyword extraction task and text preprocessing, such as: POS and MSD taggers, stemmers, lemmatisers, stop-words lists, lexical resources like WordNet, controlled vocabularies, benchmark or monitoring datasets, and other tools or resources.

The main aim of this work is to discuss the problems of keyword extraction in under-resourced languages and as the possible solution we recommend network or graph-enabled KE methods. These methods use knowledge incorporated in the structure of network or graph to extract keywords and therefore circumvent unavailable linguistic tools required in a certain KE method development.

In the second part of this paper we will explain the concept of under-resourced languages and describe the problems that occur in KE methods for such languages. The third part of the paper will explain the general procedure of network-enabled KE methods, more precisely, SBKE method through the lenses of portability to different languages. Moreover, we provide a list of available benchmark datasets for KE

© Springer International Publishing AG 2017
A. Calì et al. (Eds.): IKC 2016, LNCS 10151, pp. 124–135, 2017.
DOI: 10.1007/978-3-319-53640-8_11

development and evaluation in order to illustrate the problem of the lack of resources. The paper ends with some concluding remarks and presentation of plans for future work.

2   Deficiencies of KE Methods for Under-Resourced Languages

Next we explain the concept of under-resourced languages in the context of text analysis and keyword extraction task, and then we describe the problems that occur in KE methods for some of the European languages which have been considered as under-resourced.

2.1   Under-Resourced Languages

Today there are more than 6900 languages in the world and only a small fraction of them is supported with the resources required for implementation of Natural Language Processing (NLP) technologies or applications [2]. Authors in [2] explained that main stream NLP is mostly concerned with languages for which large resources are available or which have suddenly become of concern because of the economic interest or political influence.

The term "under-resourced languages" was introduced by Krauwer (2003) and complemented by Berment (2004). They both define criteria to consider a particular language as under-resourced: lack of a unique writing system or stable orthography, limited presence on the web, lack of linguistic expertise, lack of electronic resources for speech and language processing, such as monolingual corpora, bilingual electronic dictionaries, transcribed speech data, pronunciation dictionaries, vocabulary lists, etc. [3, 4]. Other authors have used the terms "low-density" or "less-resourced" instead of "under-resourced" languages. Further, Berment in [4] categorizes human languages into three categories, based on their digital "readiness" or presence in cyberspace and software tools: "tau"-languages: totally-resourced languages, "mu"-languages: medium-resourced languages and "pi"-languages: under-resourced languages [4]. In addition to individual researchers, these issues are recognized as important for group of researchers, and commercial technology providers, private and corporate language technology users, language professionals and other information society stakeholders gathered in Multilingual Europe Technology Alliance (META). META network is dedicated to fostering the technological foundations of a multilingual European information society with a vision of Europe united as one single digital market and information space for Language Technology [8]. In META White Paper Series the state of language technology development is categorized into the following areas: Machine Translation, Speech Processing, Text Analysis, and Speech and Text Resources. Within these areas languages can be classified into following categories: excellent, good, moderate, fragmentary and weak/no support. The most important area for KE is Text Analysis in which the languages marked with '+' in Table 1 have the lowest support [9]. Since the META is European alliance, data presented in Table 1 are related exclusively with European languages, as well as the scope of this paper. It is important to notice that the list of systematized languages in Table 1 may not be an exhaustive list of European under-resourced languages for the area of text analysis. However, there may be additional under-resourced languages such as Bosnian or Albanian which are not listed because no relevant studies were reported.

Besides to languages that are in weak support category, there are other languages that are classified into fragmentary category and few of them in moderate. As expected, English is the only language with good support in all areas (see Table 2). Expressed in the proportions: weak supported − 30%, fragmentary supported − 50%, moderate supported: 16.66% and good supported − 3.33%.

Table 1. Cross-language comparison of European languages classified according to the areas into weak/no support category [9].

Language	Machine Translation	Speech Processing	Text Analysis	Speech and Text Resources
Bulgarian	+			
Croatian	+	+	+	
Czech	+			
Danish	+			
Estonian	+		+	
Finnish	+			
Greek	+			
Icelandic	+	+	+	+
Irish	+		+	+
Latvian	+	+	+	+
Lithuanian	+	+	+	+
Maltese	+	+	+	+
Portuguese	+			
Serbian	+		+	
Slovak	+			
Slovene	+			
Swedish	+			
Welsh	+	+	+	+

Besides META systematization, credibility and objectivity of belonging to under-resourced category are also measured with BLARK (Basic Language Resource Kit) concept. BLARK is defined as the minimal set of language resources that is necessary to do any precompetitive research and education at all [3]. It must be under 10 out of 20 in order to be considered as under-resourced language. A BLARK comprises criteria, such as: written language corpora, spoken language corpora, mono and bilingual dictionaries, terminology collections, grammars, annotation standards and tools, corpus exploration and exploitation tools, different modules (e.g. taggers, morphological analyzers, parsers, speech recognizers, text-to-speech), etc. [3].

Table 2. Cross-language comparison of European good, moderate, and fragmentary supported languages in Text Analysis area [9].

Language	Good	Moderate	Fragmentary
English	+		
Dutch		+	
French		+	
German		+	
Italian		+	
Spanish		+	
Basque			+
Bulgarian			+
Catalan			+
Czech			+
Danish			+
Finnish			+
Galician			+
Greek			+
Hungarian			+
Norwegian			+
Polish			+
Portuguese			+
Romanian			+
Slovak			+
Slovene			+

2.2   Problems in Keyword Extraction and Motivation

Information Retrieval (IR) and Natural Language Processing (NLP) experts which set their research focuses on keyword extraction task, at the ACL workshop on novel computational approaches to keyphrase extraction from 2015, detected several open problems [15]: technical term extraction using measures of neology, decompounding for keyphrase extraction (especially for German language – compound morphology), extracting social oriented keyphrase semantics from Twitter, applications to noun compounds syntax and semantic, problem of over-generation errors in automatic keyword or keyphrase extraction, which is also known problem in network-enabled methods.

Another important issue but rarely discussed in the context of KE is a lack of tools for KE method development for under-resourced languages. Although there are numerous keyword extraction methods for richer-resourced languages with remarkable performance such as methods presented in [7, 10–12] (both in supervised or unsupervised setup), in the absence of language tools it is difficult to adopt them for other languages, especially for under-resourced languages. These methods are most often developed for the English language. In other words, language scalability (portability) of these methods is limited to a particular language or language group. In order to support multilingualism, and circumvent poor portability, we propose unsupervised methods, graph- or network-enabled methods for keyword extraction. Network structure enables representation of the input text as graph or network, regardless of language. In a network representation of the input text the nodes (vertices) are unique words and the edges (links) between two nodes are established when two words share a relation (e.g. co-occur within a window).

An example of graph-based method is Selectivity-Based Keyword Extraction (SBKE) [14]. Instead of developing new tools for a language of interest, application of this method requires only tuning of various parameters which are inherent for particular language (fine tuning of parameters for candidate extraction, setting the filtering thresholds for keyword expansion, …).

3 Network-Enabled KE Concept

In a network approach, network of words is used for the representation of texts, which enables the exploration of the relationships and structural information incorporated in text very efficiently. Although there are different variations, the most common way of document modeling into graph is the representation where words are modeled by vertices (nodes) and their relations are represented by edges (links). The weight of the link is proportional to the overall co-occurrence frequencies of the corresponding word pairs within a corpus. On this basis there are various possibilities for the analysis of a network structure (topology) and we will focus on the most common – network structure of the linguistic elements themselves using co-occurrence relations. This is a basic relation, but it has shown effective results in numerous studies, such as in [5–7]. Another reason to use co-occurrence, and not any semantic or syntactic relation is the lack of language tools which could extract these relations (Fig. 1).

Figure 2 presents the generalized process for portability of network-enabled keyword extraction techniques. In the first step keyword candidates are extracted from the text. After that, candidates are filtered according to properties specific for particular method. Note that in this step various network measures can be used for rankings: closeness, degree or betweenness centrality, TextRank, etc. In the final step, candidates are ranked according to the obtained value from the used measure and used thresholds, resulting with a candidate list of keywords.

3.1 SBKE Method Portability for Under-Resourced Languages

SBKE – Selectivity-Based Keyword Extraction method is a network-enabled method for keyword extraction which consists of two phases: **(1) keyword extraction** and **(2) keyword expansion**. The node selectivity value is calculated from the weighted network as the average weight distributed on the links of a single node and is then used in the procedure of keyword candidate ranking and extraction [13, 14]. This method does not require linguistic knowledge as it is derived purely from statistical and structural information of the network, therefore it is suitable for many European under-resourced languages. The main advantage is that networks are constructed from pure

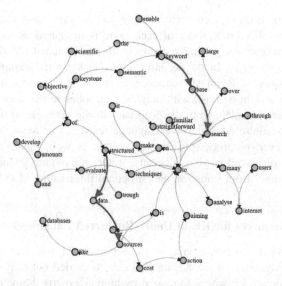

Fig. 1. Co-occurrence network constructed from text: *"KEYSTONE - semantic keyword-based search on structured data sources" is a COST Action aiming to make it straightforward to search through structured data sources like databases using the keyword-based search familiar to many internet users. The scientific objective of KEYSTONE is to analyse, design, develop and evaluate techniques to enable keyword-based search over large amounts of structured data."*

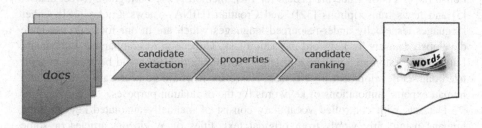

Fig. 2. Generalization of the keyword extraction techniques.

co-occurrence of words in the input texts. Moreover, the method achieves results which are above the TF-IDF (Term Frequency - Inverse Document Frequency) baseline for English and Croatian language [14].

As previously mentioned, SBKE method consist of two phases decomposed into several steps. First phase: **(1) keyword extraction**: in initial step, it is advisable that the text is preprocessed: lemmatized or stemmed (depends on tools availability for stemming or lemming in particular language). Although, preprocessing is not necessary because SBKE works without stemming or lemmatization, but it is advisable to preprocess the input text in order to reduce the size of the network, which is of importance in highly inflectional languages. After that, language network can be constructed from preprocessed input using the co-occurrence of words. For constructed network the selectivity or generalized selectivity of each node is measured as indicated in [14]. Additionally, parameters of generalized selectivity can be tuned individually for particular language or corpus. In the second phase: (2) **keyword expansion**, keyword candidates are expanded to longer sequences – two or three words long keyphrases (according to the weight of links with neighboring nodes in the network). Sequence construction is derived solely from the properties of the network. In other words, the method does not require any intensive language resources or tools except light preprocessing. However, preprocessing can be omitted as well. Finally, the method is portable to under-resourced languages because it does not require linguistic knowledge as it is derived purely from statistical and structural information of the network.

3.2   Textual Resources for KE in Under-Resourced Languages

If we want to compare the performance of the automatic KE with humans, then a valid method for the evaluation of the KE method can be carried out only by bench-mark datasets which contain keywords annotated by human experts. Some of the available datasets are presented in Table 3. It shows only those data sets that are annotated by humans (usually students involved in individual studies or human experts in a particular area). Most of the available datasets are in the English language, while other available datasets cover the French (DEFT – scientific articles published in social science journals [16]); French and Spanish (FAO 780 – FAO publications with Agrovoc terms [25, 30]); Polish (abstracts of academic papers for PKE method [17]), Portuguese (tweet dataset [5] and news transcriptions [32]), and Croatian (HINA – news articles [18]). Other languages, especially under-resourced languages which are in our focus do not have developed datasets for keyword extraction task. Collection of comparable Lithuanian, Latvian and Estonian laws and legislations (available in [19]) could be used for facilitated dataset development for KE task. However, it would be necessary to invest into human experts' annotations of keywords for the evaluation purposes.

Datasets with controlled vocabulary consist of manually annotated keywords by humans using only words from original text, titles of Wikipedia articles or some predefined list of allowed words as the controlled vocabulary. Such datasets are particularly suitable for methods which are not able to generate new words. Human annotators are also an important determinant of KE dataset - the quality of the dataset is higher if the number of human (individuals or teams) annotators is higher. Having only a single set of keywords assigned by a human annotator (individual or collaborating team) per document, taking it as the gold standard, and using the popular measures of precession, recall and their harmonic mean, F1, to evaluate the quality of keyword assigned by the automatic machine annotator ignores the highly subjective nature of key-word annotation tasks [20]. In this case Inter-Indexer Consistency (IIC) can be used instead. IIC measures the quality of keywords assigned to the test documents by developed method with those assigned by each team or human annotators.

Table 3. Available datasets with annotated keywords by human per language, number of annotators, size in the number of documents and usage of controlled vocabulary. Controlled Vocabulary is marked with yes/no when controlled vocabulary was assumed, but not always obeyed.

Language	Dataset	Controlled Vocabulary	Annotations	Num. of documents	Description
ENGLISH	SemEval2010 [22]	yes/no	authors, readers, authors and readers combined	trial: 40 training: 144 testing: 100	Student annotators from the Computer Science department of the National University of Singapore
	Wiki20 [25]	yes (Wikipedia)	15 teams	20	Computer Science papers, each annotated with at least 5 Wikipedia articles by 15 teams of indexers
	CiteULike [25]	no	330 volunteers	180	Publications crawled from CiteULike, keywords assigned by different CiteULike users who saved these publications
	FAO 30 [25, 30]	yes (thesaurus)	6 experts	30	Food and Agriculture Organization (FAO) of the United Nations publications
	500 N-KPCrowd [31]	yes	20 HITs	500 (450+50)	only the key phrases selected by at least 90% of the annotators
	Krapivin [29]	–	author assigned and editor corrected keyphrases.	2000	Scientific papers from computer science domain published by ACM
	Wan and Xiao [28]	–	-author -students	308	Documents from DUC2010, including ACM Digital Library, IEEE Xplore, Inspec and PubMed articles, author-assigned keyphrases and occasionally reader-assigned
	Nguyen and Kan [27]	–	-one by original author -one or more by student annotators	120	Computer science articles, author-assigned and reader assigned keyphrases undergraduate CS students
	INSPEC [26]	yes two sets of keywords (Inspec thesaurus) no	professional annotator	2000 training: 1000 validation: 500 testing: 500	Abstracts of journal articles present in Inspec, from disciplines Computers and Control, and Information Technology. Both the controlled terms and the uncontrolled terms may or may not be present in the abstracts

(continued)

Table 3. (*continued*)

Language	Dataset	Controlled Vocabulary	Annotations	Num. of documents	Description
	Twitter dataset [23, 24]	–	11 humans	1827 tweets training: 1000 development: 327 testing: 500	The annotations of each annotator are combined by selecting keywords that are chosen by at least 3 annotators
	Email dataset [21]	–	2 annotators	349 emails: 225 threads: 124	Email dataset consists of single and thread emails
ENGLISH FRENCH SPANISH	FAO 780 [25, 30]	yes (Agrovoc thesaurus)	-human annotator	-780 English -60 French -47 Spanish indexers working independently	FAO publications with Agrovoc terms. Documents are indexed by one indexer
POLISH	PKE [17]	yes/no	1 expert (author of the paper)	12000 training: 9000 testing: 3000	Abstracts from Polish academic papers downloaded from web sources (e.g. pubmed, yadda). All abstracts have at least 3 keywords
FRENCH	DEFT [16]	yes (50%)	author	234 training: 60% testing: 40%	French scientific articles published in social science journal
		no (50%)	students	234 training: 60% testing: 40%	
CROATIAN	HINA [18]	yes/no	8 human experts	1020 training: 960 testing: 60	Croatian news articles from the Croatian News Agency (HINA)
PORTUGUESE	Portuguese tweet dataset TKG method [5]	no	3 users	300 tweets	Portuguese tweet collections from 3 Brazilian TV shows: 'Trofeu Imprensa', 'A Fazenda' and 'Crianca Esperanca'
	110-PT-BN-KP Marujo [32]	–	-one annotator	110 news training: 100 testing: 10	The gold standard is made of 8 BN programs - 110 news subset (transcriptions), from the European Portuguese ALERT BN database

3.3 Preliminary Results

In the absence of datasets for KE in under-resourced languages (with keywords annotated by human experts or another machine algorithm), it is not possible to evaluate the SBKE method in standard measures (recall, precision, F-measure or IIC-Inter-Indexer Consistency). However, we show some preliminary results for the Serbian language. All extracted keywords from Serbian news articles available on the web portal www.novosti.rs from 3 different genres: politics, economics and sports are listed in the Table 4. It seems that SBKE method for the Serbian language prefers

Table 4. Keywords extracted from 3 different texts written on Serbian language from political, economic and sports genres.

Genre	Title	Keywords (translated to English)
POLITICS	Migrants	refugees, political, life, Angela Merkel, elections, united, more, Austria, options, Germany, population, year
ECONOMICS	Credit without a permanent job	customers, credit, capable, banks, ability, loan, institutions, interest, rates, reserve, categories, evaluation, mandatory, contract, criteria, agent
SPORTS	Serbian paralympic athletes traveled to the Rio	athlete, Rio, support, champion, medal, pride, minister, preparation, effort, table, tennis, team, London, Uroš Zeković

open-class words (such as nouns, adjectives, etc.), that are good candidates for real keywords. This was also the case for Croatian [13], and expected, since they are related Slavic languages.

4 Conclusion

This paper briefly describes graph or network-enabled keyword extraction methods. It also explains why these methods are suitable for under-resourced languages. We provide the detailed list of datasets for keyword extraction for EU languages. Using graph-based methods for keyword extraction can open the possibilities for the development of other applications which in its initial phase require keywords.

In future work we will try SBKE method for other under-resourced languages to show that knowledge incorporated in the network should replace non-existing linguistic tools necessary for keyword extraction from semi-structured web sources. In particular, we will focus on KE dataset development for Serbian, Estonian, Latvian, Lithuanian, Maltese and possibly other non-European under-resourced languages.

References

1. Beliga, S., Meštrović, A., Martinčić-Ipšić, S.: An overview of graph-based keyword extraction methods and approaches. J. Inf. Organ. Sci. **39**(1), 1–20 (2015)
2. Besacier, L., Barnard, E., Karpov, A., Schultz, T.: Automatic speech recognition for under-resourced languages: a survey. Speech Commun. **56**, 85–100 (2014)
3. Krauwer, S.: The basic language resource kit (BLARK) as the first milestone for the language resources roadmap. In: Proceedings of the 2003 International Workshop Speech and Computer SPECOM-2003, pp. 8–15. Moscow, Russia (2003)
4. Berment, V.: Méthodes pour informatiser des langues et des groupes de langues "peu dotées". Ph.D. Thesis, J. Fourier University – Grenoble I (2004)
5. Abilhoa, W.D., Castro, L.N.: A keyword extraction method from twitter messages represented as graphs. Appl. Math. Comput. **240**, 308–325 (2014)

6. Palshikar, G.K.: Keyword extraction from a single document using centrality measures. In: Ghosh, A., De, R.K., Pal, S.K. (eds.) PReMI 2007. LNCS, vol. 4815, pp. 503–510. Springer, Heidelberg (2007). doi:10.1007/978-3-540-77046-6_62
7. Mihalcea, R., Tarau, P.: TextRank: Bringing order into texts. In: Proceedings of Empirical Methods in Natural Language Processing – EMNLP 2004, pp. 404–411. ACL, Barcelona, Spain (2004)
8. META-NET – official site May 2016. http://www.meta-net.eu/
9. META-NET White Paper Series: Key Results and Cross-Language Comparison May 2016. http://www.meta-net.eu/whitepapers/key-results-and-cross-language-comparison
10. Joorabchi, A., Mahdi, A.E.: Automatic keyphrase annotation of scientific documents using Wikipedia and genetic algorithms. J. Inf. Sci. **39**(3), 410–426 (2013)
11. Lahiri, S., Choudhury, S.R., Caragea, C.: Keyword and Keyphrase Extraction Using Centrality Measures on Collocation Networks (2014). arXiv preprint arXiv:1401.6571
12. Grineva, M., Grinev, M., Lizorkin, D.: Extracting key terms from noisy and multitheme documents. In: ACM 18th conference on World Wide Web, pp. 661–670 (2009)
13. Beliga, S., Meštrović, A., Martinčić-Ipšić, S.: Toward selectivity-based keyword extraction for croatian news. In: CEUR Proceedings of the Workshop on Surfacing the Deep and the Social Web (SDSW 2014), vol. 1310, pp. 1–8, Riva del Garda, Trentino, Italy (2014)
14. Beliga, S., Meštrović, A., Martinčić-Ipšić, S.: Selectivity-based keyword extraction method. Int. J. Semant. Web Inf. Syst. (IJSWIS) **12**(3), 1–26 (2016)
15. Proceedings of the ACL 2015 Workshop on Novel Computational Approaches to Keyphrase Extraction, ACL-IJCNLP 2015, Beijing, China (2015)
16. Paroubek, P., Zweigenbaum, P., Forest, D., Grouin, C.: Indexation libreet controlee d'articles scientifiques. Presentation et resultats du defi fouille de textes DEFT2012. In: Proceedings of the DEfi Fouille de Textes 2012 Workshop, pp. 1–13 (2012)
17. Kozłowski, M.: PKE: a novel Polish keywords extraction method. Pomiary Automatyka Kontrola, R. **60**(5), 305–308 (2014)
18. Mijić, J., Dalbelo-Bašić, B., Šnajder, J.: Robust keyphrase extraction for a large-scale croatian news production system. In: Proceedings of the 7th International Conference on Formal Approaches to South Slavic and Balkan Languages, Zagreb, Croatia: Croatian Language Technologies Society, pp. 59–66 (2010)
19. Collection of comparable Lithuanian, Latvian and Estonian laws and legislations (June 2016). http://metashare.nlp.ipipan.waw.pl/metashare/repository/browse/collection-of-comparable-lithuanian-latvian-and-estonian-laws-and-legisla-tions/8d0d633eae7711e2a28e525400c0c5e f33b6cfc6ca074e1ab58859157c8374e7/#
20. Zunde, P., Dexter, M.E.: Indexing consistency and quality. Am. Documentation **20**(3), 259–267 (1969)
21. Loza, V., Lahiri, S., Mihalcea, R., Lai, P.: Building a dataset for summarization and keyword extraction from emails. In: Proceedings of the Ninth International Conference on Language Resources and Evaluation (LREC 2014). pp. 2441–2446, Reykjavik, Iceland (2014)
22. Su, N.K., Medelyan, O., Min-Yen, K., Timothy, B.: Automatic keyphrase extraction from scientific articles. Lang. Resour. Eval. **47**(3), 723–742 (2013)
23. Gimpel, K., Schneider, N., O'Connor, B., Das, D., Mills, D., Eisenstein, J., et al.: Part-of-speech tagging for twitter: annotation, features, and experiments. In: Proceedings of the 49th Annual Meeting of the Association for Computational Linguistics: Human Language Technologies: short papers – vol. 2, HLT 2011, Stroudsburg, PA, USA. Association for Computational Linguistics (2011)
24. Marujo, L., Wang, L., Trancoso, I., Dyer, C., Black, A.W., Gershman, A., et al.: Automatic keyword extraction on twitter. In: ACL (2015)

25. Medelyan, O.: Human-competitive automatic topic indexing. Ph.D. thesis. Department of Computer Science, University of Waikato, New Zealand (2009)
26. Hulth, A.: Improved automatic keyword extraction given more linguistic knowledge. In: Proceedings of the 2003 Conference on Empirical Methods in Natural Language Processing, pp. 216–223 (2003)
27. Nguyen, T.D., Kan, M.-Y.: Keyphrase extraction in scientific publications. In: Goh, D.H.-L., Cao, T.H., Sølvberg, I.T., Rasmussen, E. (eds.) ICADL 2007. LNCS, vol. 4822, pp. 317–326. Springer, Heidelberg (2007). doi:10.1007/978-3-540-77094-7_41
28. Wan, X., Xiao, J.: CollabRank: towards a collaborative approach to single-document keyphrase extraction. In: Proceedings of COLING, pp. 969–976 (2008)
29. Krapivin, M., Autaeu, A., Marchese, M.: Large dataset for keyphrase extraction. Technical Report DISI-09-055, DISI, University of Trento, Italy (2009)
30. Medelyan, O., Witten, I.H.: Domain independent automatic keyphrase indexing with small training sets. J. Am. Soc. Inf. Sci. Technol. 59(7), 1026–1040 (2008)
31. Marujo, L., Gershman, A., Carbonell, J., Frederking, R., Neto, J.P.: Supervised topical key phrase extraction of news stories using crowdsourcing. In: Light Filtering and Co-reference Normalization. Proceedings of LREC 2012 (2012)
32. Marujo, L., Viveiros, M., Neto, J.P.: Keyphrase cloud generation of broadcast news. In: Proceeding of 12th Annual Conference of the International Speech Communication Association, Interspeech (2011)

Documents and Information Retrieval

Making Sense of Citations

Xenia Koulouri[1]([✉]), Claudia Ifrim[2], Manolis Wallace[1], and Florin Pop[2]

[1] ⿆ Knowledge and Uncertainty Research Laboratory,
Department of Informatics and Telecommunications,
University of the Peloponnese, 22 131 Tripolis, Greece
{xenia.koulouri,wallace}@uop.gr
[2] Computer Science Department, Faculty of Automatic Control and Computers,
University Politehnica of Bucharest, 060042 Bucharest, Romania
claudia.ifrim@hpc.pub.ro, florin.pop@cs.pub.ro
http://gav.uop.gr
http://acs.pub.ro/

Abstract. To this day the analysis of citations has been aimed mainly to the exploration of different ways to count them, such as the total count, the h-index or the s-index, in order to quantify a researcher's overall contribution and impact. In this work we show how the consideration of the structured metadata that accompany citations, such as the publication outlet in which they have appeared, can lead to a considerably more insightful understanding of the ways in which a researcher has impacted the work of others.

Keywords: Research impact · Citations · Publication medium

1 Introduction

Academia is competitive by nature. Researchers strive to make the greatest breakthrough, attract the most lucrative funds, obtain the highest awards and achieve the greatest impact. More so now than before, this effort is not related just to excellent accomplishment but even to the very survival of the researcher in academia. Of course, this keep achieving or be ignored, or "publish or perish" as it is usually referred to, approach has its downside. The more a researcher's output is linked to their survival, the more bias they will have in assessing and presenting their work [1].

As a result, we have now reached a point where publications, once the best indicator of the value of a researcher's work [2], need to be examined with a grain of salt [3,4]. Publication records can be skewed in size [5], by publishing multiple similar papers, by splitting one work in multiple incremental publications, by exchanging gratuitous co-authorships, by repeating the same work with different datasets [6] etc., as well as in direction, by carefully selecting titles and masterfully penning abstracts to highlight relevance to one scientific field or another [7].

© Springer International Publishing AG 2017
A. Calì et al. (Eds.): IKC 2016, LNCS 10151, pp. 139–149, 2017.
DOI: 10.1007/978-3-319-53640-8_12

It is then only normal that we look not only at the publications of researchers, but increasingly also at their impact, as shown by their citations [8]. Of course, citations can also be skewed [9]. In fact, it has already been discussed that the way citations are currently examined is not sufficient [10]. In this paper we look deeper into citations, taking advantage of citations' metadata in order to achieve a better understanding and quantification of researchers' impact. Specifically, we focus on the publication medium in order to best estimate the fields of science that each work impacts.

A paper discussing similar ideas but focusing mainly on the visualization of the results has been presented at the 9th International Workshop on Semantic and Social Media Adaptation and Personalization [7]. A broader paper incorporating some of the ideas of the current work but focusing mainly on the presentation of an integrated working system is currently under consideration for publication in a special issue on "Keyword Search in Big Data" in the LNCS Transactions on Computational Collective Intelligence journal.

The remainder of this paper is organized as follows: In Sect. 2 we discuss existing approaches to the assessment and quantification of scientific impact. Continuing, in Sect. 3 we discuss the types of information that can be mined from citation metadata and in Sect. 4 we present a comprehensive methodology that uses this notion in order to achieve a deeper insight of the way in which each work and each researcher impacts the scientific world. Finally, in Sect. 5 we present and discuss some indicative results from the application of our approach and in Sect. 6 we list our concluding remarks.

2   Counting Citations

To this day citations are used to assess scientific impact. There are of course inherent weaknesses [11]; it is possible revolutionary works to go un-noticed due to random shifts of research trends or less deserving works to receive attention simply because of an inspired title [12]. Still, the fact that they are fully quantitative measures that can be computed in an automated manner with little or no human intervention makes them the measure of choice for the estimation of scientific impact.

Thus, a paper's impact is quantified as the count of citations it has received from the day it was published and up to the day of examination. This, of course, favors papers that were published many years ago, as they have been accumulating citations for a longer period of time. This is not necessarily a weakness of the measure; it is only natural that works that have been around for a longer period of time have had the opportunity to have a greater impact on the works of others. Besides, it has been observed that the yearly count of citations received by a paper diminishes after a few years; so, after some time, the advantage of earlier papers is diminished.

Similar ideas are applied towards the evaluation of the scientific value of a publication medium, such as a journal, magazine or conference. There is, though, an important difference originating in the way to use the results of this evaluation. Journals are not evaluated in order to assess which one has had the greatest overall impact on the scientific world. To the contrary, the goal is to assess the probability that an article published in a journal will make an impact in the future; readers consider this evaluation to select the journals to read and more importantly authors consider it in order to select the journals to submit to, thus maximizing the potential of their work. Therefore, the number of years that a journal has been publishing, or even the number of volumes per year or the number of articles per volume cannot be allowed to affect the evaluation.

The impact factor (IF) is the most trusted quantification of a journal's scientific potential. It is computed as the average count of citations articles published in the journal receive in the first two years after their publication; some limitations apply regarding the sources of these citations. It is clear to see that the IF is configured in a way that favors journals that publish carefully selected high quality articles, which is in accordance with the goals of journal evaluation. Of course, the impact factor is also an imperfect measure [13] and efforts are made to improve it [14].

When it comes to researchers, their past impact, and by extension their future potential, is also assessed based on citations. The first, most common and straightforward approach is the consideration of the cumulative number of citations an author has received for the complete list of their published work.

But given the highly competitive nature of the scientific community, it is rather expected that the prime tool to assess and compare researchers has received a lot of attention, both in the form of criticism of its objectivity and in the form of attempts to affect its outcomes. Numerous weaknesses have been identified, related to the number of years of activity, the effect of cooperation networks, self-citations, outlier works, frequency of publication etc.

In order to deal with the weaknesses of the count of citations as a metric, a long list of more elaborate metrics have been proposed, including the average number of citations per paper, the average number of citations per author, the average number of citations per year, the h-index [15] and similar indices [16–18], the g-index [19], the e-index [20], the s-index [21], the i-10 index, and more.

3   Citation Context

The count of citations, as well as all the other aforementioned measures that are based on it, provide a numerical quantification of impact, without any indication of where that impact has been made. This does not align well with the purpose of assessing a researcher's impact. When researchers are evaluated, for example for an academic position, only relevant publications from their publication record are considered. Still, when it comes to impact, we use the overall citation count without examining which publications they have derived from or which scientific fields they show impact in. Clearly, it would be useful to have access to such information.

In this work we examine the scientific scope of the referencing papers in order to see which fields of science have been affected by a given paper. Our goal is to describe a way to mine more information from citation records, without loosing the objectivity of the citation count, i.e. the fact that it is not directly affected by the examined researchers and it is computed with minimal user intervention. The practical question here of course is which of the citations' metadata to use and how in order to identify the scientific scope. In developing our approach we should, of course, also consider the availability of the data that will be examined.

In existing systems papers are indexed by their titles and journals they are published in; authors are indexed by the papers they have published and the keywords they use to characterize their own research interests [22]. But such metadata (titles, journals to submit to, keywords, abstract) are determined by the authors based on a priori preferences and not all are necessarily closely related to the a posteriori information regarding the actual areas that their work has an actual impact on, or even to the content of the work itself. Titles can be misleading; keywords are useful but are not standardized and are not used in all publications; textual analysis is not yet mature enough to guarantee reliable results when applied on abstracts that may be related to literary any given scientific field. More importantly, all of the above can be severely skewed by the authors, especially when they need to build a profile that shows strength in a specific field.

In contrast, the publication medium can provide a good indication of the scientific scope. When a paper is considered, either by a journal or by a conference, thematic relevance is examined together with its scientific quality. Therefore, the editorial process guarantees that, for example, papers published in the IKC conference are additionally related to semantic keyword-based search on structured data sources. Almost all edited publications come with clearly defined scopes and lists of relevant topics, and for those that do not it is relatively easy to produce them manually since this would need to be done only once for each publication medium and not separately for each article. Therefore, the automated and objective (i.e. without considering the subjective opinion of a human expert examining the specific article) consideration of the scientific scope of a given published paper is feasible.

Our approach is to examine each citation's publication medium in order to estimate the scientific field in which it indicates impact and to use this information in order to classify citations to fields and transform the unidirectional citation count - and by extension all similar metrics - into a field by field analysis which will provide much deeper insight in the way a researcher's work has impacted the rest of the scientific world.

4   Methodology

As we have already explained, our analysis is based on the examination of the publication medium. In the next paragraphs we outline the main steps required to put this notion in practice.

4.1   Preparatory Steps

The preparatory steps involve the establishment of the knowledge base that is required for the execution of the processing steps, as follows:

1. Develop a list of thematic areas
2. Compile a list of publication media (journals, magazines, conferences)
3. Assign thematic areas to each publication medium

Thematic Areas. The list of scientific fields is almost static. Therefore a reasonable first step is to acquire this hierarchy. Existing hierarchies exist that may be considered as a basis, as for example the one found in [23].

Publication Media. We can use, for example, DBLP metadata in order to acquire a first list of previous and running journals and conferences, knowing that although this list is long it is far from complete. A comprehensive list of publication media is not easy to establish. Moreover, the list is not static as some conferences disappear whilst new ones appear every year; there are similar changes to the list of journals, but they are less frequent and thus easier to tackle.

Therefore the pre-processing step regarding the acquisition of publication media is not meant to produce a complete and finalized list but rather to facilitate the initiation processing steps by dealing with the problem of cold start.

Medium to Area Assignments. Although the DBLP metadata are carefully curated, they do not contain semantic information regarding the thematic scope of the included publication media, other than their title. This title is often, but not always, enough to have a rough idea of the thematic coverage.

In order to overcome this a semi-automatic approach is needed. When we can not define the thematic area by the title of publication media, we search manual to publications' media site to determine its topic.

4.2   Processing Steps, for Each Work

For each considered article, we need to examine the list of citations as follows:

1. Acquire the list of citations
2. Identify the thematic area of each citing work
3. Aggregate findings

List of Citations. We can use Google Scholar or any other similar system to acquire a comprehensive list of citations for each article that we examine. Of course such systems are neither complete nor perfect (they inherently contain false positives, incorrectly assigned fields, damaged titles, repetitions etc.). Still, although error rates are high (often exceeding 20%), the deviation is small. Thus citations retrieved from systems such as Google Scholar are a relatively reliable source given that the error rate is similar for different articles and authors [24].

Thematic Area of Each Citation. In earlier sections we have explained that we will use the publication medium to identify the thematic scope, we have developed the lists of publication media and scopes and established the associations between publication media and thematic areas.

Thus, the connection between citing article and its thematic areas is quite straightforward.

Aggregated Impact for Each Work. Conventionally all citations associated with a published work are considered equally and uniformly, and overall impact is given as the count of citations. Given the additional thematic information that now becomes available, a rising question is the validity of considering uniformly references that have been published in a publication medium with an impact on a single science and references whose publication medium influences more than a single science. Our approach is a variable weighting factor for the two cases. In case that the papers influence a single scientific field weight will be equal to 1, whereas the weight will be distributed uniformly when multiple fields are impacted. For example, if one paper influence in two difference thematic areas, the weight will be equal to 0.5 for each scientific field. In this, all citations are considered equally as having total weighting equal to 1.

The aggregated impact for each work is given as the sum of weights, for each scientific field; as expected the impact is not calculated as a single number but rather as an array of numbers, one per field.

4.3   Processing Steps, for Each Author

For each considered article, we examine the list of public works as follows:

1. Acquire the list of published works
2. Identify the impact of each work
3. Aggregate results

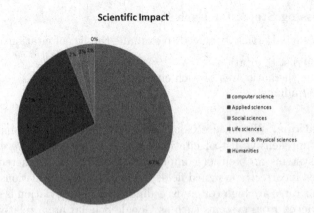

Fig. 1. Anagnostopoulos - Scientific impact

In the conventional approach, an author's citation count is calculated as the sum of citations for all of the author's published works. By extension, in our work we calculate an author's impact in each field as the sum of the impact values for that field for all of the author's works. Thus, the aggregated impact for the author is a vector calculated as the sum of the impact vectors of all of the author's published works, as calculated above.

5   Experimental Results

In order to better explain what type of insight we are looking at, in this section we examine what our approach brings to light when applied for three specific researchers, namely Prof. Ioannis Anagnostopoulos, Prof. Costas Vassilakis and Prof. George Lepouras. The results have been produced using an early software implementation of the notions presented earlier herein [25,26] (Fig. 1).

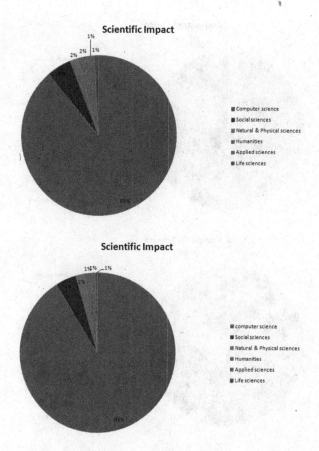

Fig. 2. Lepouras and Vassilakis - Scientific impact

5.1   Ioannis Anagnostopoulos

Ioannis Anagnostopoulos is a member of the Department of Computer Science and Biomedical Informatics, at the University of Thessaly. The position in which he has been elected faculty member, his expertise as presented in his CV and his research interests as presented in his personal web page are focused in the analysis of social networks.

Naturally, one would expect the impact of his work to be in the same area. Still, our analysis finds that 27%, of his research impact does not even lie in the field of computer science.

Looking into the details of the researcher's publications and citations we find that Prof. Anagnostopoulos worked in the fields of neural networks and image processing at the beginning of his career and much of his citation record comes from citations to work of that era. And whilst in examining Prof. Anagnostopoulos's CV we would quickly filter out these publications when evaluating him for

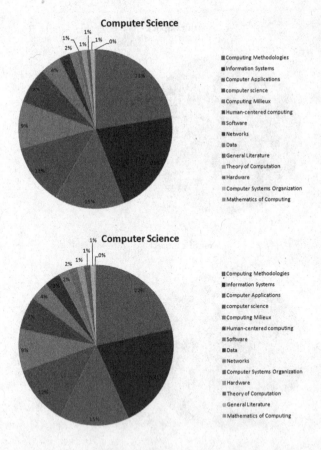

Fig. 3. Lepouras and Vassilakis - Scientific impact in computer science

his current position, using the conventional approach we would not have been able to similarly filter the 27% of his citations that are not relevant.

5.2   George Lepouras Costas Vassilakis

George Lepouras and Costas Vassilakis are members of the Department of Informatics and Telecommunications, at the University of Peloponnese. Prof. Lepouras's area of research, as indicated by position in which he has been elected faculty member, his expertise as presented in his CV and his personal statement in his personal web page lie in the field of human computer interaction. In similar fashion we can see that Prof. Vasilakis's area of research lies in the field of information systems.

Clearly, the two researchers have quite distinct works. Yet, our impact analysis shows not only that they have impact in the same broader scientific areas but also that they have very similar impact when examining detailed subfields of computer science. Whilst in the conventional approach we would consider their impact to lie in distinct areas, and more specifically in the areas that they state as their fields of expertise, our closer analysis of their citation records reveals that this would not have been accurate (Figs. 2 and 3).

6   Conclusions

In this paper we explored the information that can be extracted from citation records' metadata. In order to avoid subjectivity in the estimation and quantification of the impact we have opted to avoid author defined parameters and have instead focused our analysis on the journal or conference where a citing article has been published. This provides an objective and reliable indication of thematic scope, which allows us to see, in a semi-automated manner, which scientific areas have been affected by an author's work. The proposed approach correspond to a new and fairer evaluation of researchers.

Through some real life examples we have shown that the approach proposed herein is indeed able to provide a deeper insight into the ways in which a researcher, or even a specific paper, has impacted the scientific work. This can allow for a more fair consideration of citations in the comparative evaluation of researchers, by considering only citations belonging to the specific scientific field, as is also done for publications. Additionally, since our approach effectively partitions citations into thematic areas everyone can know the influence of each researcher and which scientific fields it represents. Also, any and all conventional citation metrics can still be applied on top of it; for example it is easy to see how to calculate the h-index per thematic area.

Of course our work is not complete. We have only just scratched the surface of the treasures hidden in the metadata of citation records. Moving forward, it would be interesting to examine to further detail the different types of impact that can be defined based on the distribution of the areas of impact [7] or to explore how impact can be redefined or refined by considering not only one but multiple hops in the citation graph.

Acknowledgments. This work has been partially supported by COST Action IC1302: Semantic keyword-based search on structured data sources (KEYSTONE).

References

1. Neill, U.S.: Publish or perish, but at what cost? J. Clin. Invest. **118**(7), 2368 (2008). doi:10.1172/JCI36371
2. Steinpreis, R.E., Anders, K.A., Ritzke, D.: The impact of gender on the review of the curricula vitae of job applicants. Sex Roles **41**(7/8), 509 (1999)
3. Fanelli, D.: Do pressures to publish increase scientists' bias? An empirical support from US states data. In: Scalas, E. (ed.) PLoS ONE **5**(4) (2010). doi:10.1371/journal.pone.0010271
4. Song, F., Parekh, S., Hooper, L., Loke, Y.K., Ryder, J., Sutton, A.J., Hing, C., Kwok, C.S., Pang, C., Harvey, I.: Dissemination and publication of research findings: an updated review of related biases. Health Technol. Assess. **14**(8) (2010). doi:10.3310/hta14080.20181324
5. Broad, W.: The publishing game: getting more for less. Science **211**(4487), 1137–1139 (1981). doi:10.1126/science.7008199
6. Kumar, M.N.: A review of the types of scientific misconduct in biomedical research. J. Acad. Ethics **6**(3), 211–228 (2008). doi:10.1007/s10805-008-9068-6
7. Wallace, M.: Extracting and visualizing research impact semantics. In: Proceedings of the 9th International Workshop on Semantic and Social Media Adaptation and Personalization, Corfu (2014)
8. van Wesel, M.: Evaluation by citation: trends in publication behavior, evaluation criteria, and the strive for high impact publications. Sci. Eng. Ethics **22**(1), 199–225 (2016). doi:10.1007/s11948-015-9638-0
9. McNab, S.M.: Skewed bibliographic references: some causes and effects. CBE Views **22**, 183–185 (1999)
10. Moed, H.F.: Citation Analysis in Research Evaluation. Springer, Dordrecht (2005)
11. Figa-Talamanca, A.: Strengths and weaknesses of citation indices and impact factors. In: Quality Assessment in Higher Education, pp. 83–88 (2007)
12. Letchford, A., Moat, H.S., Preis, T.: The advantage of short paper titles. R. Soc. Open Sci. (2015). The Royal Society. http://dx.doi.org/10.1098/rsos.150266
13. Colquhoun, D.: Challenging the tyranny of impact factors. Nature **423**(6939), 479 (2003)
14. Mutz, R., Daniel, H.-D.: Skewed citation distributions and bias factors: solutions to two core problems with the journal impact factor. J. Inf. **6**(2), 169–176 (2012)
15. Hirsch, J.E.: An index to quantify an individual's scientific research output. Proc. Natl. Acad. Sci. U.S.A. **102**(46), 16569–16572 (2005)
16. Batista, P.D., Campiteli, M.G., Konouchi, O., Martinez, A.S.: Is it possible to compare researchers with different scientific interests? Scientometrics **68**(1), 179–189 (2006)
17. von Bohlen und Halbach, O.: How to judge a book by its cover? How useful are bibliometric indices for the evaluation of "scientific quality" or "scientific productivity"? Ann. Anat. **193**(3), 191–196 (2011)
18. Bornmann, L., Mutz, R., Daniel, H.D.: The h index research output measurement: two approaches to enhance its accuracy. J. Inf. **4**(3), 407–414 (2010)
19. Egghe, L.: Theory and practise of the g-index. Scientometrics **69**(1), 131–152 (2013)
20. Zhang, C.T.: The e-index, complementing the h-index for excess citations. PLoS ONE **5**(5) (2009)

21. Silagadze, Z.K.: Citation entropy and research impact estimation. Acta Physica Polonica B **41**, 2325–2333 (2009)
22. Ifrim, C., Pop, F., Mocanu, M., Cristea, V.: AgileDBLP: a search-based mobile application for structured digital libraries. In: Cardoso, J., Guerra, F., Houben, G.-J., Pinto, A.M., Velegrakis, Y. (eds.) KEYSTONE 2015. LNCS, vol. 9398, pp. 88–93. Springer, Heidelberg (2015). doi:10.1007/978-3-319-27932-9_8
23. Glanzel, W., Schubert, A.: A new classification scheme of science fields and subfields designed for scientometric evaluation purposes. Scientometrics **56**(3), 357–367 (2003)
24. Koulouri, X.: Estimation of the area of scientific impact through the analysis of citation records, MSc thesis, Knowledge and Uncertainty Research Laboratory, University of the Peloponnese (2016)
25. Babetas, N.: Automation of the research impact estimation process, BSc thesis, Knowledge and Uncertainty Research Laboratory, University of the Peloponnese (2015)
26. Dimitriou, G.: Research area estimation and visualization, BSc thesis, Knowledge and Uncertainty Research Laboratory, University of the Peloponnese (2015)

An Ontology-Based Approach to Information Retrieval

Ana Meštrović[1] and Andrea Calì[2][(✉)]

[1] University of Rijeka, Rijeka, Croatia
amestrovic@uniri.hr
[2] Birkbeck, University of London, London, UK
andrea@dcs.bbk.ac.uk

Abstract. We define a general framework for ontology-based information retrieval (IR). In our approach, document and query expansion rely on a base taxonomy that is extracted from a lexical database or a Linked Data set (e.g. WordNet, Wiktionary etc.). Each term from a document or query is modelled as a vector of base concepts from the base taxonomy. We define a set of mapping functions which map multiple ontological layers (dimensions) onto the base taxonomy. This way, each concept from the included ontologies can also be represented as a vector of base concepts from the base taxonomy. We propose a general weighting schema which is used for the vector space model. Our framework can therefore take into account various lexical and semantic relations between terms and concepts (e.g. synonymy, hierarchy, meronymy, antonymy, geo-proximity, etc.). This allows us to avoid certain vocabulary problems (e.g. synonymy, polysemy) as well as to reduce the vector size in the IR tasks.

1 Introduction

The development of semantic technologies and of techniques for ontology management in recent years has had a significant impact on information retrieval (IR) systems. A proper use of ontologies allows to overcome certain limitations of classical keyword-based IR techniques.

There have been various approaches to IR that make use of semantic information. Early attempts to incorporate semantic knowledge into IR relied on thesauri [1,6] or WordNet [7,15]. Approaches to ontology-based IR were introduced in the late Nineties (see [9]) and continued in various directions. In this paper we propose a variant of the vector space model (VSM) originally proposed in [10].

In [16] the authors adapt the generalized vector space model (GVSM) with taxonomic relationships; in particular, they use Linked Open Data (LOD) resources as well as underlying ontologies to determine a correlation between related index terms. Castells et al. [3] adapt classical VSM with the ability to work with concepts instead of terms. Various query expansion techniques based on ontologies have been proposed in the literature [4,8,12]. The idea is to augment set of query terms with new set of semantically related concepts. A similar

© Springer International Publishing AG 2017
A. Calì et al. (Eds.): IKC 2016, LNCS 10151, pp. 150–156, 2017.
DOI: 10.1007/978-3-319-53640-8_13

approach can be applied on the document set as well. In [2] document expansion based on graphs is proposed. While document expansion approaches manage to resolve some vocabulary problems (and thus increase precision and recall), one of its drawbacks is the redundancy of information.

Dragoni et al. [5] propose another approach for document expansion which manages to partially solve the redundancy problem: they use only a limited set of *base terms* (extracted from WordNet) for document descriptions. This approach is based on the WordNet ontology and inclusion relation among terms. The idea of [5] we adopt and extend this idea of using limited set of base terms for document and query representation as in [5]. Our approach is generalised to multiple ontological *layers*[1]. Each layer is an ontology of its own and can incorporate any relation, e.g. synonymy, term similarities, semantic relation, geo-proximity etc.

The main goal driving our research is to extend the set of possible relations from inclusion relation to other possible lexical and semantic relations. This way we are able to avoid certain vocabulary problems and to reduce the number of dimensions of the vector space model.

Our contribution can be summarised as follows.

1. We adopt the vector space model and we introduce a *base taxonomy* that is extracted from an ontology that can be a lexical database such as Word-Net, Wiktionary or any thesaurus. Documents and queries are represented as vectors of *base concepts* from the base taxonomy.
2. We extend the framework with the possibility of incorporating new ontologies represented as graphs, thus incorporating semantic relations at different levels of abstraction into the framework. Notice that we envision the inclusion of ontologies suitably extracted from the Linked Data cloud which, despite their flat nature in terms of logical data representation, contain semantic information at different levels of abstraction.
3. We define a set of *mapping functions* which map multiple ontological layers onto the base taxonomy. This way, each concept from the incorporated ontologies can be represented as a vector of base concepts from the base taxonomy as well.
4. We propose a general weighting scheme, which is used for the document ranking task.

2   IR with Base Taxonomy

In our framework we use an ontology-based vector space model (VSM) for the representation of both documents and queries. We now briefly recall the VSM. The main idea in the Vector Space Model is that each document

---

[1] The different ontology layers are not actually layered, strictly speaking, but they constitute different aspects of ontological information that can be somewhat seen as layers.

and query is a *point* in a n-dimensional space. For a given set of k documents $D = \{D_1, D_2, \ldots, D_k\}$ and query q, a document D_i is represented as a vector $\mathbf{D}_i = \langle w_{i1}, w_{i2}, \ldots, w_{in} \rangle$; the query is also represented as a vector $\mathbf{q} = \langle w_{q1}, w_{q2}, \ldots, w_{qn} \rangle$. In the classic keyword-based VSM, each dimension correspond to one term or keyword from the document set. Weights may be determined by using various weighting schemes, among which the most widely adopted, in keyword-based VSM, is the *tf-idf*. Following the approach defined in [5], we adopt a VSM where each term is expressed in terms of a set of *base concepts* $\mathbf{B} = \{b_1, \ldots, b_\ell\}$. In particular, in [5] the set $\mathbf{B}$ is extracted from the WordNet database as the set of all concepts that do not have hyponyms (that is, roughly speaking, leaf concepts in the generalisation taxonomy).

The concept weight is calculated from the WordNet graph $T = \langle N, A \rangle$ by taking into account all *explicit* and implicit *occurrences* of a term b_1 of a *base vector* $\mathbf{B} = \langle b_1, \ldots, b_\ell \rangle$. An implicit occurrence of b_i is computed as the "projection" of an ancestor c of b_i on b_i. The concepts $b_1, \ldots, b_\ell$ are *leaves* (or more generally nodes without successors in the graph) in T, and an occurrence of any ancestor c of some b_i constitutes an implicit occurrence of b_i with an *information value* u_i (with $i \in \{1, \ldots, \ell\}$) computed as follows: if $\langle c = c_0, c_1, \ldots, c_h, b_i \rangle$ is the path in T from c to b_i, u_i is defined as

$$u_i(c) = \prod_{j=0}^{h} \frac{1}{fanout(c_j)} \tag{1}$$

where $fanout(c_j)$ denotes the number of children of c_j in T. If c is not an ancestor of b_i, naturally $u_i(c) = 0$; if c is an element of $\mathbf{B}$, say $c = d_j$, then $u_i(d_j) = 1$.

Example 1. Consider the graph T as depicted in Fig. 1. The base vector is $\mathbf{B} = \langle b_1, b_2, b_3, b_4 \rangle$ and the information vector for c_1 is $\mathbf{u}(c_1) = \langle 1/3, 1/6.1/6, 1/3 \rangle$, while $\mathbf{u}(c_2) = \langle 0, 1/2, 1/2, 0 \rangle$. ∎

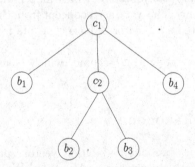

Fig. 1. Taxonomy graph T for Examples 1 and 2.

Once the information vector $\boldsymbol{u}(c) = \langle u_1, \ldots, u_\ell \rangle$ is defined for each concept c, the document vector representation $\mathbf{D} = \langle d_1, \ldots, d_\ell \rangle$ is computed by defining its elements as

$$w_i = \sum_{c \in N} u_i(c) \cdot occ(c) \tag{2}$$

where $occ(c)$ is the number of occurrences (or better of lexicalisations) of c in the document, and N, we remind, is the set of nodes of the graph T.

Example 2. Consider again the reference taxonomy of Example 1. If the document D is $b_4c_1c_1b_2c_1c_2b_4$, we have $occ(c_1) = 3$, $occ(c_2) = 1$, $occ(b_2) = 1$ and $occ(b_4) = 2$. Therefore the document representation vector $\mathbf{D}$ is

$$\begin{aligned}
\mathbf{D} &= 3\mathbf{u}(c_1) + 1\mathbf{u}(c_2) + 1\mathbf{u}(b_2) + 2\mathbf{u}(b_4) = \\
&\quad 3\langle 1/3, 1/6, 1/6, 1/3 \rangle + \langle 0, 1/2, 1/2, 0 \rangle + \langle 0, 1, 0, 0 \rangle + 2\langle 0, 0, 0, 1 \rangle = \\
&\quad \langle 1, 2, 1, 5/2 \rangle.
\end{aligned}$$

∎

The document ranking is carried out by using cosine similarity. For a given document D_i and a query q cosine similarity is defined as

$$sim(\mathbf{D}_i, \mathbf{q}) = \frac{\mathbf{D}_i \cdot \mathbf{q}}{\|\mathbf{D}_i\| \cdot \|\mathbf{q}\|}. \tag{3}$$

This approach allows us to work with the base concepts only, while the impact of non-base terms (which are more general concepts in the ontology T) is taken into account by "diluting" their contribution as described above.

In the next section we propose a generalisation of this approach which takes into account different ontologies at once, thus offering the possibility of carrying out information retrieval tasks under different types of ontological information.

3   Ontology-Based Framework

In this section we present our general framework for ontology-based information retrieval. Our aim is to offer the possibility of incorporating ontological information of various types into the approach described in Sect. 2.

3.1   Formalisation of the Taxonomy/Base Ontology

In [5] a linear combination of base concepts is defined according to the inclusion (generalisation) relation extracted from WordNet, as explained in the previous section. This approach allows for expressing terms on the basis of a smaller term base, with an evident advantage in terms of efficiency. However, notice that such approach considers only the generalisation, while several other semantic relations among terms can be considered. In this paper we extend the approach of [5] to a more general framework that can capture *any lexical relation* (synonymy, near synonymy, meronymy etc.). We make use of the *reference ontology* T, which we represent as a directed graph $T = \langle N, A \rangle$ – notice that T does not need to be a tree/hierarchy, that is, it is not necessarily a taxonomy with a generalisation relation. The reference ontology T can be the WordNet graph (with generalisation arcs) considered by [5] and summarised in Sect. 2, but other ontologies can be employed.

First of all we notice that, formally, we can represent the information vector **u** of the previous section as the values imposed by all nodes in N onto those of **B** through a function $f : N^2 \rightarrow [0,1]$; in particular, $f(c_1, c_2)$ is the value imposed on c_2 by one occurrence of c_1 as described in Sect. 2 (see Eq. 1). Therefore we can rewrite Eq. 2 as

$$d_i = \sum_{c \in N} f(c, b_i) \cdot occ(c). \tag{4}$$

Since additional semantic information can be stored in other ontologies external to T, we formalise the additional knowledge as a set of semantic relations $\{E_1, \ldots, E_m\}$; each E_i, where $E_i \subseteq E$, represents a specific property in the ontology (e.g. lexical proximity, concept similarity etc.); in other words, we can have several types of arcs (m types, specifically) on V, each with its own semantics. Notice that the overall relation E is merely the union of all the relations $E_1, \ldots, E_m$, which in general may come from different ontological sources. In fact this setting can be suitably (and obviously) represented as a set of ontological graphs $G_1 = \langle N, E_1 \rangle, \ldots, G_m = \langle N, E_m \rangle$, all with the same set of nodes N as the reference graph T (this is a limitation which we shall remove later).

Now we define a set of functions $F = \{f_1, f_2, \ldots, f_m\}$, with f_i defined as $f_i : N^2 \rightarrow [0,1]$, for $i = 1, \ldots, m$. Each function defines how the information weights spread according to the relation (ontology) E_i; in particular, $f_i(c_1, c_2)$ defines the influence on c_2 of the occurrence of c_1, similarly to what we have described in Sect. 2.

Example 3. Consider again the graph of Example 2. Suppose we have an occurrence, in a part-whole relation E_k (with $1 \leqslant k \leqslant m$), that c_1 is part of some concept c_3 (for instance, *toe* is part of *foot*). Assume then that, similarly to what we illustrated in Sect. 2, we "spread" information across E_k from the whole to the part by dividing the weight equally (and recursively) into the parts. So, if c_3 has, say, 4 parts, the weight of an occurrence of c_3 will split into its parts and therefore $f_k(c_3, c_1) = 1/4$. For other relations E_j different from E_k, the function f_j could be completely different; for instance, if E_j is a (symmetric) geographic proximity (or adjacency) relation, $f_j(c_1, c_2)$ could have non-zero value only if c_1 and c_2 are adjacent. ∎

Now, the $E_1, \ldots, E_m$ provide information to the document representation by means of the base vector similarly to what we have presented in Sect. 2. In particular, we can write an extension of Eq. 4 as follows.

$$d_i = \sum_{c \in N} f_i(c, b_i) \cdot occ(c) + \sum_{j=1}^{m} \sum_{c \in N} f_j(c, d_i) \cdot occ(c) \tag{5}$$

for all i such that $1 \leqslant i \leqslant \ell$.

Notice that external ontologies may have terms not appearing in T, which is realistic, given that specialised ontologies might include terms which are not even in the Wordnet or DBpedia graph. In such a case it is sufficient, given an ontology graph $G = \langle V, E \rangle$, to define a function $g : V \times N \to [0, 1]$, where $g(c_1, c_2)$ defines the influence of c_1 (in G but not necessarily in T) on c_2 (in T). This of course, from the practical point of view, requires a non-trivial mapping of the nodes of G to those of T, which might be a hard challenge.

Example 4. Consider a geographic *proximity* relation represented by a graph $E = \langle N, E \rangle$. Let $d(n_1, n_2)$ be the distance between two nodes n_1, n_2 of N in E. We could assign $f(n_1, n_2) = 1/2$ if $d(n_1, n_2) = 1$; $f(n_1, n_2) = 1/4$ if $d(n_1, n_2) = 2$ and $f(n_1, n_2) = 1/4$ if $d(n_1, n_2) > 2$. This would model a propagation of the proximity up to two steps in the proximity graph, with associated weighting. ∎

4   Discussion

In this paper we presented an ontological approach to Information Retrieval, extending the framework of [5], which utilises a *base vector* for the VSM model, thus reducing the vector size and increasing efficiency. We are able to incorporate ontological information of any kind, rather than restricting ourselves to taxonomies (generalisation hierarchies). We defined functions to incorporate ontological information in a base vector in a formal way.

This work has still a long way to go to validate the approach. While we are aware of suitable ontologies to be incorporated into the framework, the effectiveness of the approach needs to be verified through a careful experimentation, which we plan to perform. Moreover, the problem of dealing with ontologies that include terms not present in the reference ontology requires a careful investigation, as defining the influence of terms not appearing in the reference ontology seems to be a hard challenge[2].

Acknowledgments. This research was funded by the COST Action IC1302 *semantic KEYword-based Search on sTructured data sOurcEs* (KEYSTONE).

References

1. Aronson, A.R., Rindflesch, T.C., Browne, A.C.: Exploiting a large thesaurus for information retrieval. In: RIAO, vol. 94 (1994)
2. Baziz, M., et al.: An information retrieval driven by ontology from query to document expansion. In: Large Scale Semantic Access to Content (Text, Image, Video, and Sound). LE CENTRE DE HAUTES ETUDES INTERNATIONALES D'INFORMATIQUE DOCUMENTAIRE (2007)
3. Castells, P., Fernandez, M., Vallet, D.: An adaptation of the vector-space model for ontology-based information retrieval. IEEE Trans. Knowl. Data Eng. **19**(2), 261–272 (2007)

---

[2] A simplistic approach could be to ignore the influence for such terms, but we do believe a more suitable technique ought to be devised.

4. Carpineto, C., Romano, G.: A survey of automatic query expansion in information retrieval. ACM Comput. Surv. (CSUR) **44**(1), 1 (2012)
5. Dragoni, M., da Costa Pereira, C., Tettamanzi, A.G.B.: A conceptual representation of documents and queries for information retrieval systems by using light ontologies. Expert Syst. Appl. **39**(12), 10376–10388 (2012)
6. Hersh, W.R., Greenes, R.A.: SAPHIRE–an information retrieval system featuring concept matching, automatic indexing, probabilistic retrieval, and hierarchical relationships. Comput. Biomed. Res. **23**(5), 410–425 (1990)
7. Mandala, R., Tokunaga T., and Tanaka H.: The use of WordNet in information retrieval. In: Proceedings of the Conference on Use of WordNet in Natural Language Processing Systems (1998)
8. Navigli, R., Velardi, P.: An analysis of ontology-based query expansion strategies. In: Proceedings of the 14th European Conference on Machine Learning, Workshop on Adaptive Text Extraction and Mining, Cavtat-Dubrovnik, Croatia (2003)
9. Luke, S., Lee S., Rager, D.: Ontology-based knowledge discovery on the world-wide web. In: Working Notes of the Workshop on Internet-Based Information Systems at the 13th National Conference on Artificial Intelligence (AAAI 1996) (1996)
10. Salton, G., Wong, A., Yang, C.-S.: A vector space model for automatic indexing. Commun. ACM **18**(11), 613–620 (1975)
11. Schuhmacher, M., Ponzetto, S.P.: Knowledge-based graph document modeling. In: Proceedings of the 7th ACM International Conference on Web Search and Data Mining. ACM (2014)
12. Song, M., Song, I.Y., Hu, X., Allen, R.B.: Integration of association rules and ontologies for semantic query expansion. Data Knowl. Eng. **63**(1), 63–75 (2007)
13. Thomopoulos, R., Buche, P., Haemmerlé, O.: Representation of weakly structured imprecise data for fuzzy querying. Fuzzy Sets Syst. **140**(1), 111–128 (2003)
14. Tsatsaronis, G., Panagiotopoulou, V.: A generalized vector space model for text retrieval based on semantic relatedness. In: Proceedings of the 12th Conference of the European Chapter of the Association for Computational Linguistics: Student Research Workshop. Association for Computational Linguistics (2009)
15. Voorhees, E.M.: Query expansion using lexical-semantic relations. In: Croft, B.W., van Rijsbergen, C.J. (eds.) SIGIR 1994. Springer, London (1994)
16. Waitelonis, J., Exeler, C., Sack, H.: Linked data enabled generalized vector space model to improve document retrieval. In: NLP and DBpedia Workshop, ISWC 2015, Bethlehem, 11–15th September 2015
17. Wong, S.K.M., Ziarko, W., Wong, P.C.N.: Generalized vector spaces model in information retrieval. In: Proceedings of the 8th Annual International ACM SIGIR Conference on Research and Development in Information Retrieval. ACM (1985)

Collaboration and Semantics

Game with a Purpose for Verification
of Mappings Between Wikipedia and WordNet

Tomasz Boiński[(✉)]

Department of Computer Architecture, Faculty of Electronics,
Telecommunication and Informatics,
Gdańsk University of Technology, Gdańsk, Poland
tobo@eti.pg.gda.pl

Abstract. The paper presents a Game with a Purpose for verification of
automatically generated mappings focusing on mappings between Word-
Net synsets and Wikipedia articles. General description of idea standing
behind the games with the purpose is given. Description of TGame system, a 2D platform mobile game with verification process included in the
game-play, is provided. Additional mechanisms for anti-cheating, increasing player's motivation and gathering feedback are also presented. The
evaluation of proposed solution and future work is also described.

1 Introduction

In 2012 Samsung Electronics Polska performed a study among people in Poland
on the time spend on video games [2]. Almost half of the population aged 27–35
spends 1 to 2 h weekly playing games and 14% spends over 20 h a week. High
percentage of the players use mobile devices like smartphones (20%) and tablets
(5%). It can be seen that in many cases playing games occupies the amount of
time equal to at least a part-time job. On the other hand many nowadays problems still cannot be solved by a computer algorithm and finding human resources
to perform such tasks is difficult. A simple example of such problem is image
tagging or verification of results obtained by traditional heuristics. Design of a
proper solution linking those two areas could prove to be useful for performing
laborious tasks without a need of hiring additional workers.

Some types of problems, namely numerically solvable ones, adopted volunteer
computing model [1] where the users donate the power of their machines when it
is not needed (the calculations are done between the periods of active hardware
usage). Using this model it is not possible to solve all type of problems, as some
of them cannot be successfully turned into a computer algorithm [12]. We can use
heuristics but than we still have to verify the results manual. In crowdsourcing [3]
approach the user is encouraged to perform a task for some type of gratitude.
The task can be both algorithmic and non-algorithmic.

In this paper we focus on a third model, so called human-based computation
(HBC) [17]. It is using human brain directly to solve a computational problem.
The term was defined by Kosorukoff in 2001 in a paper about human enhanced
genetic algorithm [5].

© Springer International Publishing AG 2017
A. Calì et al. (Eds.): IKC 2016, LNCS 10151, pp. 159–170, 2017.
DOI: 10.1007/978-3-319-53640-8_14

HBC can be viewed as similar to crowdsourcing. The later focuses solely on solving the problem by human, while in the former model part of the problem is solved by a computer. Usually the machine performs sub-problem organization, distribution and retrieval of results, sometimes some calculations are done using heuristics. The human part usually contains the verification of computer generated results or performing the calculations itself [8]. The question arises what if we could use all potential resources (time and knowledge of the users and hardware capabilities of their devices) to solve non-algorithmic problems? One can imagine that if we would treat a group of users as a distributed system, then it is sufficient to divide a problem into small portions, distribute them to the players and finally aggregate achieved results. This however introduces some difficulties, from technical ones like how to divide a problem into sub-problems, how to distribute them and how to gather results, to more social oriented like how to trust the results and more importantly how to convince people to spend their time on solving our problem.

In our research we aimed to apply HBC-model for verification of mappings. We developed our solution for the purpose of mappings verification between WordNet and Wikipedia [4,9–11]. The mappings were obtained during Colabmap project and are a result of running heuristics on a computer. Currently we are working on implementation of the optimal client for linking data and a generalization of our solution hoping to provide a general framework suitable for solving different types of problems.

2   Games with a Purpose

In 2006 Luis von Ahn proposed usage of computer games as something more than pure entertainment and thus creating the idea of so called GWAP (Game With A Purpose) [12]. GWAPs are typical games that provide standard entertainment value that users expect but are designed in a way that allows generation of added value by solving a problem requiring intellectual activity. It is worth noticing that GWAPs does not allow financial gratification for the work. The will to continue playing should be treated as the only way of gratifying users [14].

Ahn defined three types of GWAPs:

- output-agreement game,
- inversion-problem game,
- input-agreement game.

In the first type of GWAPs two randomly selected players are presented with identical input data and each produces results only based on the available information. Both players should achieve identical results without any knowledge about the other player – they are awarded only when both will give identical answers. In this case an identical answer provided by both players is treated as highly probable to be correct as it comes from independent sources. Example of such game is The ESP Game [13], where users task was to tag images with keywords. The players were presented with an image and were given 2,5 min to enter all keywords that are related to the image. The game proved to be very popular. During first few months authors managed to gather around 10 million tags with statistics showing many users playing for over 40 h a week [12]. In 2006 Google released their own version of the game called Google Image Labeler[1] (it was shut down in 2011) which was used to extend capabilities of Google Graphics.

The second type, the inversion-problem game, also selects players randomly. The players are however divided into two groups – describers and guessers. The describer is presented with input data and is responsible for creating tips allowing the guessing player to correctly point out the input data. The players are awarded points when the output given by the guesser is equal to the input. One of the examples of such game is Phetch [15]. One of the players is presented with a random image from the Internet. His or her task is to describe the image to other players. Other players task is to find an identical image. Other example is the Peekaboom game [16]. The task of the players is to quest words that are describing an image. The "boom" player is presented with an image and its description in a few words. The "peek" player is presented with empty page on which the "boom" player gradually reveals parts of the image. The "peek" player have to guess, based on the revealed fragments, the exact words describing the image.

The third type, input agreement game, also selects players randomly. Both players have to achieve agreement on the input data. More precisely they have to guess whether the other player received the same or different input data. Each player describes what he or she sees on the screen and the other player have to state whether the input is similar to theirs or different. The example of such game is TagATune [6] where players should describe their feelings about a tune that is played. Based on the description the players have to decide whether they heard the same or different tune.

What is common for all above types of games is that the players unknowingly generate added value that is not possible to calculate using computers. The problem behind such games is a way to lure players – only very large user base can provide viable results. During implementation many techniques can be used to enrich the game and encourage more players, like time limits, awards in form of points and achievements, difficulty levels, leader boards or randomness of input data [14].

The quality of target game can be described by two parameters: average lifetime play (the time that average player spent playing the game) and throughput (average number of problems solved per hour of playtime) [12]. Simko [7] also pointed out that GWAP should be characterized also by the total number of players that took part in the game.

3   Wikipedia - WordNet Connections

3.1   Colabmap Project

Large text repositories requires efficient methods for information retrieval. To improve retrieval a background linguistic knowledge should be provided.

---

[1] http://en.wikipedia.org/wiki/Google_Image_Labeler.

This knowledge can be represented in many ways: ontologies, semantic networks, controlled vocabularies. There is lack of tools allowing to use them together. During the Colabmap project [4,9–11] we created a set of mappings between English Wikipedia articles and WordNet synsets.

It is worth noting that not all WordNet synsets can be mapped to Wikipedia articles. Often general terms are not present in Wikipedia. For instance synset friend (a person you know well and regard with affection and trust) cannot be found in Wikipedia. The closest match we could find was article friendship. In the project we were less interested in vague matches and we are looking for exact matches. We preferred not to create a mapping than create a wrong one. For that reason we valued accuracy over coverage. The accuracy has been measured as a percent of correctly mapped synsets to all mapped synsets.

Four different algorithms were used for mappings creation:

- The unique results algorithm – based on the fact that most of WordNet phrases are used in one synset only. If a phrase is unique and querying Wikipedia returns only one result then we create a mapping. Such an observation allowed us to find related articles for 32,232 synsets which is 39% of all synsets. The evaluation for 100 random synsets has revealed an accuracy of 97%. That gives us 32,024 mapped synsets out of 82,115 total synsets.
- The synonyms algorithm – there are 21.4% Wikipedia articles with redirects and 49% WordNet synsets have synonyms. We assumed that if the same article occurs at least twice in the results from querying Wikipedia with synonym words from WordNet then a mapping exists. The synonyms algorithm has covered 22% of synsets with 88% accuracy. That gives us 18,065 mapped synsets, 15,897 of which are correct. Harvard, Harvard University [a university in Massachusetts] is an example where the algorithm works well. Querying Wikipedia with the Harvard phrase gives us 14 results whereas Harvard University 13 results. Both queries return the Harvard University article at the top position in the result set, thus it is recognized as the correct one.
- Exact matches algorithm – this approach creates a mapping whenever an article title and a synset phrase are the same, but only if the phrase is used in no more than one synset. As a result 59% of synsets have been matched with articles with a measured accuracy of 83%. That gives us 48,447 synsets, 40,211 of which are correct. The strength of this algorithm lies in the fact that 51% of synsets have exactly one sense and define such unique terms as Lycopodium obscurum, Centaurea, Green Revolution etc. There are some wrong mappings e.g.: the fishbone [a bone of a fish] synset, which is mapped to the Fishbone [Fishbone is a U.S. alternative rock band formed in 1979 in Los Angeles].
- Most used algorithm – this approach was based on an assumption that the first returned result from the Wikipedia Opensearch API is the correct one. If a synset has synonyms, then we select an article that appears the most frequently and at the highest positions among all returned results. This very simple approach was tailored for improving the overall coverage. However, it has introduced a very high number of wrong mappings. As many as 84% synsets have been mapped with a measured accuracy of only 17%. That gives us 68,976 mapped synsets, but with only 11,726 correct.

Final mapping was an intersection of the Unique Results, Synonyms and Exact matches algorithms, which have produced the best results. The algorithms have been run separately and results merged. In the effect 60,623 synsets were mapped, which is 74% of all noun synsets with a measured accuracy of 73%, which is as many as 44,254 correctly mapped synsets. Sample mappings are presented in Table 1. Each mapping consists of a WordNet synset, definition of the synset and the title of Wikipedia article with special characters coded using RFC 3986. Such mappings, when proved to be correct, will allow formalization of Wikipedia structure. The problem here is that without further verification we have no knowledge which mappings are in fact correct.

3.2   Validation

Tempted by the results obtained during the Samsung's survey we decided on implementing a GWAP for validation of those connections. The originally obtained mappings were extended with three additional "next best" mappings with the idea of presenting the user a question (definition of a synset) with 4 possible answers (Wikipedia article titles). At the beginning the 3 other answers were randomly selected from the set of Wikipedia's pages but such approach quickly proved to be incorrect as the "next best" mappings were not related at all to the question. Instead we used Wikipedia search functionality to select alternative answers (according to Wikipedia) following the algorithm:

for all synonyms of WordNet synset **do**
  Read the synonym
  Perform a Wikipedia search using the synonym
  Store 3 top elements from search results
end for

The example of extended mappings are presented in Table 2. For the tests we randomly selected 100 synsets from our database to limit the time needed to gather the results and verify the viability of the game.

4   TGame

We decided to implement TGame[2] ("Tagger Game") as a 2D platform game following the output-agreement model. We chose Android platform as a test environment due to its popularity and ease of access for users and developers. The game implements standard features like different levels and collectibles (coins, hearts), need of finishing one level before the other one is accessible. The player is encouraged to replay levels by a simple point system that awards the player for killing monster, gathering stars and hearts (Fig. 1).

---

[2] https://play.google.com/store/apps/details?id=pl.gda.eti.kask.tgame, http://kask. eti.pg.gda.pl/tgame/.

Table 1. Sample WordNet – Wikipedia mappings

Synset (WordNet)	Definition (WordNet)	Article (Wikipedia)
Sept. 11, September 11, 9–11, 9/11, Sep 11	The day in 2001 when Arab suicide bombers hijacked United States airliners and used them as bombs	September_11
Interval, time interval	A definite length of time marked off by two instants	Time
Ice age, glacial epoch, glacial period	Any period of time during which glaciers covered a large part of the earth's surface	Ice_age, Glacial_period
Man hour, person hour	A time unit used in industry for measuring work	Man-hour
Entity	That which is perceived or known or inferred to have its own distinct existence (living or nonliving)	Entity
French leave	An abrupt and unannounced departure (without saying farewell)	French_leave
Hunt, hunting	The pursuit and killing or capture of wild animals regarded as a sport	Huntingdon
Blindman's bluff, blindman's buff	A children's game in which a blindfolded player tries to catch and Identify other players	Blind_man%27s_bluff
Landler	A moderately slow Austrian country dance in triple time	L%C3%A4ndler
Coup d'oeil, glimpse, glance	A quick look	Coup_d%27%C5%93il

4.1   Tying Questions with the Game

One of the biggest challenge is to properly include the mappings into the game. We tried to implement the questions to be as non intrusive as possible but still easy to stumble on. In TGame the verification of mappings is done when the player wants to activate a checkpoint (a respawn place when player is moved when killed). To activate the checkpoint player needs to answer the question provided by marking the correct mapping (Fig. 2). When the answer is identical to the one stored in the database the checkpoint is activated. If the player chose other answer then the checkpoint is not enabled. When the player is certain that he or she selected a correct answer then he or she can report his or her answer using the proper option in the pause menu. The checkpoint is then activated for one use.

From the technical point of view the communication between the client and the server goes as follows. Each client upon first connection downloads pack of configurable number of questions and possible answers thus internet connection is required only at first start of application and later at user chosen intervals.

Table 2. Sample of extended mappings, the original mapping is emphasized

Synset (WordNet)	Articles (Wikipedia)
Sept. 11, September 11, 9–11, 9/11, Sep 11	*September 11*, 9/11 Commission, 9/11 conspiracy theories, United Airlines Flight 93
Interval, time interval	*Time*, Interval (music), Interval, Interval (mathematics)
Ice age, glacial epoch, glacial period	*Ice age*, Pleistocene, Wisconsin glaciation, Gravettian
Man hour, person hour	*Man-hour*, Hourman, Man of the Hour, 24 h of Le Mans
Entity	*Entity*, Administrative divisions of Mexico, Administrative division
French leave	*French leave*, French leave (disambiguation), Desertion, French Leave (1930 film)
Hunt, hunting	*Huntingdon*, Hunting, Fox hunting, Seal hunting

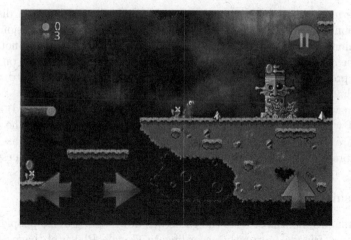

Fig. 1. TGame

Whenever possible the game sends gathered results with statistical information and downloads new pack of questions (if needed).

4.2   Answer Verification

The process of reporting wrong answers requires explicit action from the user. It was designed to require some activity but not too much so not to discourage the users. Very easy access to submission would encourage people to skip reading the question and just submitting information about wrong answer. In general the game has to be paused and proper menu have to be selected. Only the last question can be reported.

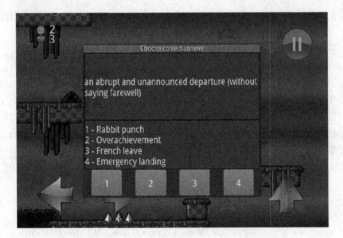

Fig. 2. Checkpoint activation

Furthermore when submitting results also time elapsed between displaying the question and selecting the answer is also submitted. Such extensions allows us to eliminate submissions that i.e. are so short that the user would not be able to read the question. Randomly selected batch of questions required on average 5 s to be properly read and understood by players. We decided to discard all answers that took less than 4 s (8% of all results).

The answers that the players gave (correct, incorrect and reported) are later compared with the one calculated by Colabmap algorithms. All the selected answers are visible in administrators panel. Upon clicking on phrase ID the admin is presented with all answers that were submitted for given question.

4.3   Results Analysis

During first two months of tests the game was downloaded by 25 people, mainly students and friends (currently according to Google Play web page there are between 50 and 100 downloads without any additional advertising). The original 25 players gave 626 answers for 100 questions. The game run 10 h in total on multiple devices. Each player solved 44,42 questions per hour. At this rate, with average playtime of each player at 5 h, we would need minimum 2500 players to answer each question at least once. Judging by other similar games available on Google Play such number can be achieved with proper advertising of the product given the user base and popularity of mobile games.

During the evaluation of the proposed solution we faced two main type of problems. In some cases the additional answers generated using the Wikipedia search functionality provided very similar pages which introduced difficulty in choosing the correct one. Selection of 100 random questions also introduced problem with high variety of difficulty level among questions. It became obvious that some of them require expert-level knowledge. Examples of questions belonging to those two groups are presented in Table 3. The column "Answer 1" contains the correct mapping. Furthermore the type of game implemented (a simple platform game) did not match the questions asked.

As a result for our future study we are developing a mechanism to asses difficulty of a given question. We are also developing a different type of client that will better suite such type of question (a quiz game).

5 Mapping Update

We tried three approaches for deciding whether the mapping, based on the answers provided by the players, should be updated or not:

- The mapping was considered correct when 75% of the player answers agreed. This approach however did not give any results as only 50% of original mappings managed to get enough answers, none of the incorrect mappings were marked as correct.
- The mapping was considered correct when at least 50% of player answers agreed. In this case 64% of all mappings were marked as correct which covered 75% of all mappings marked as correct in our database. Unfortunately this method generated some false positives.
- The mapping which gathered the most of the player answers was considered correct. In this case 74% of all mappings were marked as correct witch covered 80% of all mappings originally marked as correct in our database. This method also generates false positives.

Currently in our solution we implemented the third method as it provided the best results. Still this method does not allow us to automatically state whether the given mapping is correct or not. However "problematic" mappings are clearly pointed out by the players. Such mappings can than be verified manually by

Table 3. Sample questions with high level of difficulty

The question	Answer 1	Answer 2	Answer 3	Answer 4
Asiatic nut trees: wing nuts	Pterocarya	Pterocarya fraxinifolia	Pterocarya stenoptera	Cyclocarya
A colorless flammable volatile liquid hydrocarbon used as a solvent	Xylene	O-Xylene	P-Xylene	Xylene cyanol
A former large county in northern England	Yorkshire	Yorkshire 6	Yorkshire captaincy affair of 1927	South Yorkshire Fire and Rescue
Fine porcelain that contains bone ash	Bone China	Aynsley China	Bisque (pottery)	Porcelain

Table 4. Updated mappings

WordNet definition	Original mappings	Other available answers
An advanced student or graduate in medicine gaining supervised practical experience (houseman' is a British term)	Internet Movie Database	Houseman, Julius Houseman, *Internship (medicine)*
Large voracious aquatic reptile having a long snout with massive jaws and sharp teeth and a body covered with bony plates	Crocodile tears	*Crocodile*, Crocodylus, Schnappi
(Elections) more than half of the votes	Supermajority	*Majority*, Simple majority, Absolute majority

experts. In our further work we plan on extending the procedure with additional parameters like user reputation, level and field of education, history of answers etc. which should improve the level of trust that can be put in user the generated answers.

During the evaluation period the players submitted 17 mappings update requests regarding 12 questions. Sample reports are presented in Table 4. The corrected mappings are emphasized.

6   Conclusion

We proposed a platform for verification of the results of heuristic algorithms. Currently verification of mappings is supported. We verified the solution using Wikipedia – WordNet mappings and managed to get some promising results and were able to correct some of the mappings. The problems that still need to be solved include better formulation of the question and the trust that the system can put in answers provided by the users.

We also plan on extending the proposed solution by generalizing it for other type of tasks, inclusion of different clients, not only game based, designed for certain types of questions or with required user knowledge in mind. We are also currently implementing social features like achievements and leader boards that should lure more players and create a wider user base. In case of Wikipedia - WordNet mappings we plan on tagging questions with difficulty levels and include them in a quiz-like game similar to "Fifteen to One"[3] or "1 of 10"[4]. Such type of client could be more suitable for such defined problems. The TGame can be a great application for crowd base image tagging or a client when the questions will be redesigned to a Yes/No format.

Our research shows that popularity of computer games and wide availability of devices that can be used for playing at any time makes GWAPs an approach

---

[3] http://en.wikipedia.org/wiki/Fifteen_to_One.

[4] http://pl.wikipedia.org/wiki/Jeden_z_dziesi%C4%99ciu.

that has some unexplored potential. Our first implementation, despite its draw-back, shows that this potential is relatively easy to unlock. Even for small user base we managed to find some errors in our mappings. Implementation of different client applications more fitting the types of tasks needed to be done (image tagging, sound analysis etc.) and careful question formulation should enable us to fully unlock the possibility of crowdsourcing based task execution. When succeeded such possibility can be of great help to researchers around the world as it reduces resources and time needed to verify the results of designed algorithms and implementations. Furthermore it can be implemented as an alternative to in app purchases or advertisements. This way the users can be provided with content with their work be treated as another means to "pay" for it.

References

1. Anderson, D.P., Fedak, G.: The computational and storage potential of volunteer computing. In: Sixth IEEE International Symposium on Cluster Computing and the Grid, CCGRID 2006, vol. 1, pp. 73–80. IEEE (2006)
2. Biuro Prasowe Samsung Electronics Polska Sp. z o.o.: Prawie połowa Polaków gra codziennie w gry wideo (in Polish)
3. Howe, J.: Crowdsourcing: a definition, p. 29 (2006). http://www.crowdsourcing.com/cs/2006/06/crowdsourcing_a.html
4. Korytkowski, R., Szymanski, J.: Collaborative approach to WordNet and Wikipedia integration. In: The Second International Conference on Advanced Collaborative Networks, Systems and Applications, COLLA, pp. 23–28 (2012)
5. Kosorukoff, A.: Human based genetic algorithm. In: 2001 IEEE International Conference on Systems, Man, and Cybernetics, vol. 5, pp. 3464–3469. IEEE (2001)
6. Law, E.L., Von Ahn, L., Dannenberg, R.B., Crawford, M.: TagATune: a game for music and sound annotation. In: ISMIR, vol. 3, p. 2 (2007)
7. Simko, J.: Semantics discovery via human computation games. In: Sheth, A. (ed.) Semantic Web: Ontology and Knowledge Base Enabled Tools, Services, and Applications, p. 286. IGI Global, Hershey (2013)
8. Simko, J., Bieliková, M.: Games with a purpose: user generated valid metadata for personal archives. In: 2011 Sixth International Workshop on Semantic Media Adaptation and Personalization (SMAP), pp. 45–50. IEEE (2011)
9. Szymański, J.: Mining relations between Wikipedia categories. In: Zavoral, F., Yaghob, J., Pichappan, P., El-Qawasmeh, E. (eds.) NDT 2010. CCIS, vol. 88, pp. 248–255. Springer, Heidelberg (2010). doi:10.1007/978-3-642-14306-9_25
10. Szymański, J.: Words context analysis for improvement of information retrieval. In: Nguyen, N.-T., Hoang, K., Jędrzejowicz, P. (eds.) ICCCI 2012. LNCS (LNAI), vol. 7653, pp. 318–325. Springer, Heidelberg (2012). doi:10.1007/978-3-642-34630-9_33
11. Szymański, J., Duch, W.: Self organizing maps for visualization of categories. In: Huang, T., Zeng, Z., Li, C., Leung, C.S. (eds.) ICONIP 2012. LNCS, vol. 7663, pp. 160–167. Springer, Heidelberg (2012). doi:10.1007/978-3-642-34475-6_20
12. Von Ahn, L.: Games with a purpose. Computer 39(6), 92–94 (2006)
13. Von Ahn, L., Dabbish, L.: Labeling images with a computer game. In: Proceedings of the SIGCHI Conference on Human Factors in Computing Systems, pp. 319–326. ACM (2004)
14. Von Ahn, L., Dabbish, L.: Designing games with a purpose. Commun. ACM 51(8), 58–67 (2008)

15. Von Ahn, L., Ginosar, S., Kedia, M., Blum, M.: Improving image search with phetch. In: IEEE International Conference on Acoustics, Speech and Signal Processing, ICASSP 2007, vol. 4, p. IV-1209. IEEE (2007)
16. Von Ahn, L., Liu, R., Blum, M.: Peekaboom: a game for locating objects in images. In: Proceedings of the SIGCHI Conference on Human Factors in Computing Systems, pp. 55–64. ACM (2006)
17. Wightman, D.: Crowdsourcing human-based computation. In: Proceedings of the 6th Nordic Conference on Human-Computer Interaction: Extending Boundaries, pp. 551–560. ACM (2010)

TB-Structure: Collective Intelligence for Exploratory Keyword Search

Vagan Terziyan[1], Mariia Golovianko[2(✉)], and Michael Cochez[1,3,4]

[1] Faculty of Information Technology, University of Jyvaskyla,
Jyvaskyla, Finland
{vagan.terziyan,michael.cochez}@jyu.fi
[2] Department of Artificial Intelligence, Kharkiv National University
of Radioelectronics, Kharkiv, Ukraine
mariia.golovianko@nure.ua
[3] Fraunhofer Institute for Applied Information Technology FIT,
Sankt Augustin, Germany
[4] Informatik 5, RWTH Aachen University, Aachen, Germany

Abstract. In this paper we address an exploratory search challenge by presenting a new (structure-driven) collaborative filtering technique. The aim is to increase search effectiveness by predicting implicit seeker's intents at an early stage of the search process. This is achieved by uncovering behavioral patterns within large datasets of preserved collective search experience. We apply a specific tree-based data structure called a TB (There-and-Back) structure for compact storage of search history in the form of merged query trails – sequences of queries approaching iteratively a seeker's goal. The organization of TB-structures allows inferring new implicit trails for the prediction of a seeker's intents. We used experiments to demonstrate both: the storage compactness and inference potential of the proposed structure.

Keywords: Keyword search · Query trail · TB-structure · Collective intelligence

1 Introduction

The collection of extremely large volumes of digital information having emerged during the last decades referred usually to as big data is highly promising: the expected impact of insights derived from large data sets is broadly recognized [1]. It has been shown that big data analytics gives deeper understanding of processes and helps recognizing hidden patterns exposing radically new knowledge which can be translated into significantly improved decisions and performance across industries. Thus big data is considered a new type of asset that brings a new culture of informed data-driven decision making.

Simultaneously, big data handling is exceptionally challenging due to its volume, velocity and variety [2–4]. This boosts the problem of big data accessibility implying therefore the need of new data models and techniques for big data storage and retrieval. Despite the fact that essential results have already been achieved to build search

© Springer International Publishing AG 2017
A. Calì et al. (Eds.): IKC 2016, LNCS 10151, pp. 171–178, 2017.
DOI: 10.1007/978-3-319-53640-8_15

engines, there is still an obvious need for new approaches to large-scale search. Among others, Marz et al. [5], Cambazoglu and Baeza-Yates [6] focus on the scalability problem related to the high computational costs of storing and processing large volumes of distributed data, Lewandowski [7] deals with providing fast access to large amounts of information and effectiveness providing relevant search results and optimizing the search process for a user.

This paper represents a new collaborative filtering technique used for exploratory search in big data. It is based on the assumption that a search engine can recognize real intents and information needs of a new seeker faster or more accurately than can be done by the users themselves. Previous search experience of users in the form of users' queries organized in a special compact way will help to approach the intended search goal, with a minimal number of iterations. The main contribution of the research reported in this paper is the development of a tree-based data structure which is used for both compact storage of collective search experience and prediction of seekers' real information needs at an early stage of their search activity.

2 Related Work

The shift of the monopoly of the classical information retrieval to the concept of iterative and interactive search performed by search engines has been studied recently [8]. While information retrieval methods work well for closed world bases when the search output can be well predicted at the beginning of the search process, it is obviously insufficient when seekers have scarce knowledge of a topic, cannot specify a goal and a correct query to represent their information needs promptly. This is typical for search in open world systems, which are rapidly accumulating large-scale data and knowledge which is never complete. It was noticed that people's conceptions of their information problems change through the interactions with the dynamically changing environment containing other humans, sensors, and computational tools [9]. The search becomes more an *exploration* than a *retrieval*.

Traditional *Information retrieval* is best served by analytical strategies engaging in simple lookup activities: sending a carefully planned series of queries posed with precise syntax to a search engine and getting search results in the form of discrete and well-structured objects with minimal need for further examination and comparison. Effectiveness and efficiency of information retrieval is ensured by traditional techniques of crawling, indexing and ranking [10] engaging various combinations of syntactical, statistical and semantic analysis of either content itself or its structure and experimenting with various forms and models of page content and search query representation.

Exploratory search [11] aims at new information learning and investigation, i.e., new knowledge development. Such search involves multiple iterations and requires cognitive processing and interpretation of in-between search results before the search goal is achieved. A user iteratively applies a combination of querying and browsing strategies, makes on-the-fly selections and navigations, studies and assesses search results, compares and integrates the obtained information, and finally develops knowledge. In this view, search becomes a more user-dependent or user-driven process called *interactive information retrieval* [8], which is far more complicated than the usual matching of queries and documents and their further ranking. To ensure a precise query formulation a series of user-engine interactions is done: query formulation, query modification, and inspection of the list of results. The overall process is usually referred to as *query expansion* [12, 13].

In this research we propose a new exploratory search approach which takes advantage of a new query expansion technique working with a tree-based data structure for query log storage. Our idea is to go further than simple query reformulation and to use hidden patterns in users' querying history for more effective organization of big data search (e.g., web pages). Recognition of hidden models or patterns in users' behavior can not only provide a basis for multipurpose analytics, e.g., extracting the exact semantics intended by the user [14], but also contribute to essential optimization of data processing and more accurate and fast information provisioning to end users.

3   Smart Data Structure for Search Trails Organization

3.1   TB-Structures and Their Application for Search Trails Storage

Users' collective behavior keeps hidden patterns that can be used by a search engine for prediction of new users search intents at early stages of search. In our research we focus on the sequential and interactive nature of exploratory search which allows approaching a search goal by mutual co-evolution of three components: a user's knowledge and intents, queries and query answers. It is done by a series of iterations called seeking episodes – interactions of a seeker, a search engine and the content provided by the engine addressing a single search goal. A single seeking episode lasts from the first recorded time stamp to the last recorded time stamp on the search engine server from a particular searcher during a particular search period [15]. A 30-minutes inactivity timeout or the termination of the browser or the tab is usually used to demarcate seeking episodes in a web log.

A seeking episode is a set of triples (S_i, Q_i, R_i), where Q_i is a query sent by a seeker to a search engine; R_i is a set of web pages generated by a search engine in response to the query; S_i is a state of a user's knowledge after the study of R_i and processing of a subset R_i' containing the content considered relevant by a seeker. The search process is iteratively continued implying a change of seeker's knowledge and thus further query expansion. Three types of search sequences or trails show comprehensively the search dynamics: a queries trail, a response trail, and a knowledge (mind). The last one trail shows how a user's understanding of the problem has changed along with the query expansion. Analysis of all S, Q, R trails can be used for search optimization.

This paper covers the first stage of the study: query trails processing and organization for further search optimization. We assume that a seeker can benefit from collective search experience represented as organized query trails leading to satisfaction of users information needs. We call the field of computational learning dealing with algorithms of query trail organization, pattern recognition and a classifier induction "query trail learning" (QTL) and use a predictive model based on a data structure invented by Lovitskii and Terzyan [16] called a "there-and-back" structure or simply a

TB-structure. In their paper a TB structure is defined as a merge of a tree forest and an inverted tree forest. A TB structure was initially applied to combinatorial pattern matching for indexing and generating string data and proved to be useful for fast full text searches. A words space modeled with TB-structures is constructed from trees merged by the common prefixes and suffixes containing nodes denoted with alphabet symbols. A TB-structure is promising due to the possibility of its self-growth as a result of automatic generation of new links and new subtrees constructions.

We use a TB-structure for query trails storage and generation. It is a group of nodes where each node has a value in form of a keyword query: a word or a string of words sent by a seeker to a search engine, and a set of references to other nodes indicating possible transfers to next queries during search. A root layer contains nodes denoted by a set of possible initial queries, the leaves layer keeps terminal queries – the ones leading to either satisfaction (the search goal achievement) or disappointment (the search termination because of inability to reach the goal), and internal nodes are organized as intersected subtrees. Roots and leaves layers are formed by sets of non-repeating elements.

A QTB-structure allows mapping observations about a user's initiated query expansion to a target value, i.e., a query formulated in a way, so that an answer from a search engine would satisfy real information needs of a user. The structure has the power to generate new implicit relationships, thus new query trails can be automatically inferred over the basic structure. Prefixes of QTB-structures help clustering people by their initial intents and similar knowledge while suffixes define seekers' real information needs.

3.2 Algorithm of a QTB-Structure Feeding

Construction of a QTB-structure also called QTB feeding is an iterative process of incorporation of new query trails into an existing structure. Query sets are stored in form of strings $Trail_i : \{Q_{i1}, Q_{i2}, \ldots, Q_{it}\}$ in batches $Batch_k\{Trail_i : \{Q_{i1}, Q_{i2}, \ldots, Q_{it}\}\}$, where Q_{i1} is the first query in the i-search session and Q_{it} is the terminal one. A batch contains a set of all non-repeating trails captured by a search engine. The order of query trails influences a resulting QTB-structure view and its compactness.

The feeding procedure uses an incremental strategy which takes query trails out of a batch one by one and insert them into a QTB-structure QTB. QTB is traversed first top-down and then bottom-up with the purpose of finding the longest common "top-down" path $Trail_i^{td}$ from the root node that matches a prefix of the trail:

$$Trail_i^{td} : \{Seq[1,\ldots m]\}, \text{ where } Seq[1,\ldots m] \subseteq \{Trail_i\} \text{ and } Seq[1,\ldots m] \subseteq QTB,$$

and the longest common bottom-up path $Trail_i^{bu}$ from a leaf that matches a suffix of a trail for a query trail and a QTB-structure:

$$Trail_i^{bu} : \{Seq[1,\ldots n]\}, \text{ where } Seq[1,\ldots n] \subseteq \{Trail_i - Trail_i^{td}\} \text{ and } Seq[1,\ldots n] \subseteq QTB.$$

$Seq_{ij}[1]$ is always a root node while $Seq_{ij}[n]$ is a leaf. A node traversed at the j-level in a QTB must also be at the j-position in the added trail. For example, see Fig. 1, in case of the QTB-structure containing one trail $Trail_1 : \{Q_1, Q_7, Q_9, Q_{12}, Q_{18}\}$ and a new trail $Trail_2 : \{Q_1, Q_7, Q_{10}, Q_{13}, Q_{19}\}$ the longest common top-down path is $Trail_2^{td} : \{Q_1, Q_7\}$ and the longest common bottom-up path is $Trail_2^{bu} : \{\emptyset\}$.

After $Trail_i^{td}$ and $Trail_i^{bu}$ are defined, new nodes $Trail_i^{new}$ to be inserted into the QTB-structure can be obtained:

$$Trail_i^{new} : Trail_i - Trail_i^{td} - Trail_i^{bu} \text{ or } Seq_{ij}[m+1, \ldots, l-1].$$

$Trail_i^{new}$ needs to be added to QTB. We insert $Trail_i^{new}$ by linking the last node of $Trail_i^{td}$ to the first node of the sequence $Trail_i^{new}$ and the last node of $Trail_i^{new}$ to the first node of $Trail_i^{bu}$ (see Fig. 1).

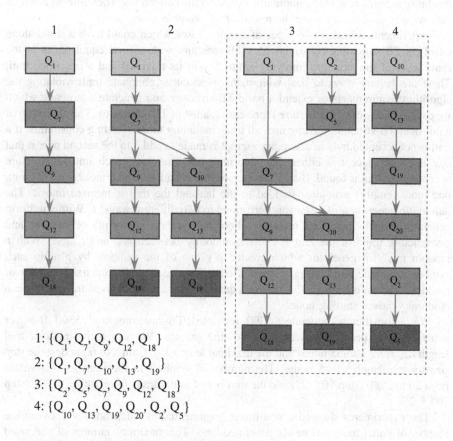

1: $\{Q_1, Q_7, Q_9, Q_{12}, Q_{18}\}$
2: $\{Q_1, Q_7, Q_{10}, Q_{13}, Q_{19}\}$
3: $\{Q_2, Q_5, Q_7, Q_9, Q_{12}, Q_{18}\}$
4: $\{Q_{10}, Q_{13}, Q_{19}, Q_{20}, Q_2, Q_5\}$

Newly generated: $\{Q_2, Q_5, Q_7, Q_{10}, Q_{13}, Q_{19}\}$

Fig. 1. An example of a QTB-structure feeding

Imagine that a new trail $Trail_4$: $\{Q_{10}, Q_{13}, Q_{19}, Q_{20}, Q_2, Q_5\}$ has to be incorporated into the structure from Fig. 1. The algorithm will reveal that $Trail_4^{td}$ and $Trail_4^{bu}$ are empty and will add a new independent trail into the structure. It makes a structure less compact and causes some duplication but eliminates the situation when cycles appear. We are planning to solve the problem of the possible ambiguity of TB-structures' traversal with swarm intelligence, ant-algorithms particularly, which allow choosing the best alternatives according to collective experience.

4 Experiments

The idea behind the first round of the experiments was (1) to reveal all possible cases causing controversial or ambiguous situations to ensure the applicability of the algorithm to real keyword queries and (2) evaluate the generative power of TB-structures: an ability to infer new trails, implicit for initial trails collections. Therefore an artificial data set was created to ensure the reveal of all possible trails.

Experiments revealed two types of specific cases which could have a stand-alone effect on TB-structures construction: a TB-structure with several equal nodes at the same level of the structure implying ambiguity in its traversal and a one node trail. There are several ways to deal with such cases: either eliminate trails violating the algorithm requirements or extend a basic TB-structure and generate a structure which we called *a fail-over TB-structure* representing a list of TB-structures. The algorithm of a fail-over TB-structure feeding uses all trails including those violating constraints: if a trail is not accepted by a structure, an attempt is made to add it to the second one, if that one does not accept it either, the algorithm continues to search until a structure accepting the trail is found. If all TB-structures are exhausted and none accepts the trail, one more (empty) structure is added to the list and the trail is inserted into it. The *automatic generation of trails* was performed in two different ways: 1. With a uniform generator, which generated trails by choosing symbols uniformly at random and constructing trails of the length chosen randomly between l_{min} and l_{max}; 2. With a random traversal generator which created a graph of the symbols by placing each symbol in a node and adding a directed edge from each node to a fixed number of randomly chosen other nodes. A trail was generated by a random graph traversal from a randomly chosen starting node.

We automatically generated 5390 restricted TB-structures and 5390 fail-over structures with permutations of the following settings. We varied the minimal trail length l_{min} from 5 to 55 nodes and the maximal length l_{max} from 5 to 60 nodes; the step size for the length was 5 nodes. The number of symbols in the alphabet was varied from 20 to 1280 (step $10 * 2^n$) and the number of initial trails from 100 to 51200 (step $100 * 2^n$).

The experiments showed a non-linear, exponential-like dependency between the number of initial trails and newly generated ones. The maximum number of generated trails was achieved when two conditions were satisfied: a big difference between the values of the minimal and the maximal possible length of trails ($l_{min} = 5; l_{max} = 60$) and the smallest possible alphabet ($alphabet = 20$). The biggest explosion of new trails was observed in a fail-over structure, which is not surprising given the fact that the fail-over structure contains at least all trails in a normal TB structure. In many cases the number of generated trails was so large that the number could not be counted within reasonable time.

The combinatorial explosion is not critical in our case because usually real search tasks imply shorter query trails than the ones we used in the experiments, and larger alphabets.

5 Conclusions and Future Work

A web seeker traditionally uses keywords to formulate search objectives to a search engine. Typical IR systems answer a query literally and return a set of results accordingly. Quite often "human-machine misunderstanding" or "misinterpretation" leads to the need of multiple iterations approaching a query which is understood similarly by a user and the search engine. Current techniques of search optimization use so called automatic query expansion (AQE) are based on semantic or syntactic similarity of terms used in queries. However open-world systems operating with big data require new search techniques due to data volumes, variety and velocity. Engines searching in open world settings should be proactive and predict hidden information needs of a seeker; becoming rather a navigator towards a real information need, not only an IR system.

We offer a technique for search optimization benefiting from collective search experience captured and processed by a search engine. Users' collective behavior contains hidden patterns that can be used by a search engine for the prediction of new users' search intents at early stages of the search process. This paper demonstrates how new forms of tree-based structures' organization can contribute to effective storage and operation over big data sequences, e.g., query trails, describing the search experience of multiple users. We build our theory on top of the idea of the sequential nature of the search process implying that a search goal is approached by mutual co-evolution of three components: person's knowledge and intents, queries and query results. This research is the first step towards comprehensive optimization of exploratory search. A tree-based TB-structure described in the paper is chosen as a smart data model used for compact storage of explicit query trails and inference of implicit trails useful for new users' intents prediction. Recognition of hidden patterns by analyzing explicit query trails and inferring new ones can be applied to the process of knowledge discovery, e.g., to support ontology learning process. And vice versa, semantic conceptualization of keyword queries based on ontologies can enhance the described search algorithm and provide additional semantics to TB-structures.

The structure itself is promising and gives wide possibilities of applications besides keyword search. It can be applied for various tasks implying sequential processes and configurations in biology, medicine, industry, academic field, for logistics and planning, etc. Experiments show that generative power of the proposed data structures is very high, in some cases we experience explosion of new implicit knowledge emergence. This is both an opportunity (due to learning capabilities of the systems built over TB-structures) and a challenge (because the bigger volumes of information cause processing complexity). In our ongoing research we are addressing this problem by application of swarm intelligence for TB-structures self-organization and prediction of a user's real information needs.

Acknowledgements. This article is based upon work from COST Action KEYSTONE IC1302, supported by COST (European Cooperation in Science and Technology).

References

1. Mayer-Schönberger, V., Cukier, K.: Big Data: A Revolution That will Transform How We Live, Work, and Think. Houghton Mifflin Harcourt, Canada (2013)
2. McAfee, A., Brynjolfsson, E., Davenport, T.H., Patil, D.J., Barton, D.: Big data. Manag. Revolution Harvard Bus Rev. **90**(10), 61–67 (2012)
3. Chen, H., Chiang, R.H., Storey, V.C.: Business intelligence and analytics: from big data to big impact. MIS Q. **36**(4), 1165–1188 (2012)
4. Chen, M., Mao, S., Liu, Y.: Big data: a survey. Mob. Netw. Appl. **19**(2), 171–209 (2014)
5. Marz, N., Warren, J.: Big Data: Principles and Best Practices of Scalable Realtime Data Systems. Manning Publications Co., New York (2015)
6. Cambazoglu, B.B., Baeza-Yates, R.: Scalability challenges in web search engines. Synth. Lect. Inf. Concept Retrieval Serv. **7**(6), 1–138 (2015)
7. Lewandowski, D.: Evaluating the retrieval effectiveness of web search engines using a representative query sample. J. Assoc. Inf. Sci. Technol. **66**(9), 1763–1775 (2015)
8. Bao, Z., Zeng, Y., Jagadish, H.V., Ling, T.W.: Exploratory keyword search with interactive input. In: Proceedings of the 2015 ACM SIGMOD International Conference on Management of Data, pp. 871–876. ACM, May 2015
9. Belkin, N.J., Cool, C., Stein, A., Thiel, U.: Cases, scripts, and information-seeking strategies: on the design of interactive information retrieval systems. Expert Syst. Appl. **9**(3), 379–395 (1995)
10. Brin, S., Page, L.: Reprint of: The anatomy of a large-scale hypertextual web search engine. Comput. Netw. **56**(18), 3825–3833 (2012)
11. Marchionini, G.: Exploratory search: from finding to understanding. Commun. ACM **49**(4), 41–46 (2006)
12. Efthimiadis, E.N.: Interactive query expansion: a user-based evaluation in a relevance feedback environment. J. Am. Soc. Inf. Sci. **51**(11), 989–1003 (2000)
13. Fattahi, R., Parirokh, M., Dayyani, M.H., Khosravi, A., Zareivenovel, M.: Effectiveness of Google keyword suggestion on users' relevance judgment: a mixed method approach to query expansion. Electron. Libr. **34**(2), 302–314 (2016)
14. Bobed, C., Trillo, R., Mena, E., Ilarri, S.: From keywords to queries: discovering the user's intended meaning. In: Chen, L., Triantafillou, P., Suel, T. (eds.) WISE 2010. LNCS, vol. 6488, pp. 190–203. Springer, Heidelberg (2010). doi:10.1007/978-3-642-17616-6_18
15. Jansen, B.J., Spink, A.: How are we searching the World Wide Web? A comparison of nine search engine transaction logs. Inf. Process. Manag. **42**(1), 248–263 (2006)
16. Lovitskii, V.A., Terziyan, V.: Words' Coding in TB-Structure. Problemy Bioniki **26**, 60–68 (1981). (In Russian)

Using Natural Language
to Search Linked Data

Viera Rozinajová and Peter Macko[✉]

Faculty of Informatics and Information Technologies,
Slovak University of Technology in Bratislava, Bratislava, Slovakia
{viera.rozinajova,peter.macko}@stuba.sk

Abstract. There are many endeavors aiming to offer users more effective ways of getting relevant information from web. One of them is represented by a concept of Linked Data, which provides interconnected data sources. But querying these types of data is difficult not only for the conventional web users but also for experts in this field. Therefore, a more comfortable way of user query would be of great value. One direction could be to allow the user to use a natural language. To make this task easier we have proposed a method for translating natural language query to SPARQL query. It is based on a sentence structure - utilizing dependencies between the words in user queries. Dependencies are used to map the query to the semantic web structure, which is in the next step translated to SPARQL query. According to our first experiments we are able to answer a significant group of user queries.

Keywords: Semantic Web · Natural language processing · Linked Data

1 Introduction

In the recent years the has become the first place where people look for information, which brings new user groups to this ecosystem. These users are not experienced computer experts but rather common people, which need to work with data on the web in an easy and straightforward way. However, data on the web (texts, videos, pictures etc.) are currently stored mainly in unstructured form, what makes searching for information very difficult.

Fortunately, there is another possibility of storing information on the web, in a fully structured way: this formalism is called Semantic Web[1] or specifically Linked Data[2]. The Semantic Web brings the possibility to store data using RDF standard and to publish them on many endpoints.

---

[1] Semantic Web - http://www.w3.org/standards/semanticweb/.
[2] Linked Data - http://linkeddata.org/.

© Springer International Publishing AG 2017
A. Calì et al. (Eds.): IKC 2016, LNCS 10151, pp. 179–189, 2017.
DOI: 10.1007/978-3-319-53640-8_16

There has already been a lot of research performed in order to transform current unstructured web to structured information form, so huge knowledge bases like Yago[3], DBpedia[4] or Freebase[5] are available. There is a lot of data in the knowledge bases so the next challenge is how to use these data effectively and how to find information quickly and in a user friendly way.

The goal of our work is to utilize the advantages of Semantic Web and at the same time to make it attractive for current web users. Nowadays, methods, that are used for searching on the web, are based on keywords (general search providers) and facets (e-shops). These methods are currently used in the area of the Semantic Web, but they do not utilize its advantages. People use relations between words in their speech, search engines which use keywords are not able to understand these relations. Façade search engines are able to understand only relations which are defined by author of engine.

A common method for communication among people is the natural language. In the natural language there are used relations between words, same as they are used between entities in the Semantic Web. Therefore, it would be advantageous to use the natural language as a way of searching for information in the Semantic Web, where the relationships between things are well described. Another part of the story is concerned with the habits of current web users: as we know, they are mostly focused on keyword search. Thus it would be useful to give them a possibility to ask questions not only in natural language but also in semi-natural or keyword form.

One example of the question can be: *Which German cities are birth places of physicists with Nobel Price?* This type of question is quite difficult even impossible to answer using only keywords. If we use knowledge base DBpedia, this question can refer to triples: *?city dbo:country dbr:Germany, ?person dbo:birthPlace ?city, ?person dbo:type yago:Physicist110428004, ?person dbo:award ?award, ?award dbo:subject ?dbr:Nobel_Price* and the answer to this question is list represented by *?city* variable. As we can see, the question can be in some points well mapped to triples, but in some cases it is necessary to find connections between the words, for example *yago: Physicist110428004* can be translated to the word *Physicist*. Some entities are hidden in the query like a connection to the *?person* variable and also the connection between the words *city* and *Germany* is not well visible. Also in this query there is no direct connection to *Nobel Price*, because it is connected by the element *dbo:Nobel_Price_in_{field}*, where parameter field could be physics, peace etc. In the next parts of the text, we look at these problems and clarify how our method deals with them.

This paper is organized as follows: In the next section we introduce methods used in the field of natural language interface. Section 3 addresses the problem, introduces our method for its solution and describes the main steps: dataset preprocessing and translation process. Section 4 presents results and the last part contains the conclusion.

---

[3] Yago - https://www.mpi-inf.mpg.de/departments/databases-and-information-systems/research/yago-naga/yago/.

[4] DBpedia - http://wiki.dbpedia.org/.

[5] Freebase - https://www.freebase.com/.

2 State of the Art

The problem of the natural language interface for database (NLIDB) [1] has been here for a relatively long time. Firstly, the problem was defined on relational databases [12], but nowadays many approaches concentrate on the Semantic Web. The main focus of all methods in this field is the translation process of the user's input to SPARQL[6] language that is de facto standard for querying the Semantic Web.

The natural language is manifold and according to [13] user's queries can be divided into seven groups, whereby the biggest groups are: list, definition and predictive question types. We focus on these types of questions in our analysis. In the next part we will describe some of the state-of-the-art solutions in this field.

The method NLION [10] is designed to understand the controlled natural language by using reformulated semantic structure. The method uses its own query analysis, where the words from query are tagged as properties or concepts. The main drawback of the method is the ability to answer only factual questions and the need of expressing by the user query full path between entities in dataset.

GiNSENG [2] and the system "Sorry, I only speak natural language" [11] are another works based on the controlled natural language, where all queries are translated by using templates. While GiNSENG uses templates that are recursively constructed, "Sorry, I only speak natural language" uses strictly defined templates. This definition depends on previous analysis of SPARQL queries. The main drawback of both works are templates for queries which have to be accepted by users.

The next group of methods, Querix [9] and Panto [15], is directed towards solving a vocabulary problem [7] which is significant especially in this area. To settle this problem, the methods use the pre-processing, in which datasets have to be analyzed, and many methods create lexicons based on words in the dataset and its alternative phrases.

The second common methodology uses the sentence parser, mainly the Stanford Parser[7], for the sentence analysis. Querix uses this parser by extracting the main word categories. By using these categories, the method identifies usable patterns. Panto, on the other hand, uses the Stanford Parser to build sentence structure and from this structure, the method uses connections between words. Panto uses the most complex method, but as well as the previously mentioned works, it needs to have specified a full path between entities. The second issue of this method is the user interface, because this method does not help user at the time of writing query. The Querix and a FREyA [5] are also focused on another significant problem, ambiguity problem, which is solved by asking a user additional questions in case of ambiguity.

User interface plays an important role in a query typing process; it shows to a user without previous experience how to start or construct a query; it can streamline the user to use the accurate words to avoid future ambiguities. The method OWLPath [14] supports suggestion by showing a list of possible words and the user has to select from this list. GiNSENG [2] uses another approach which shows the possible continuation

---

[6] SPARQL Protocol and RDF Query Language - http://www.w3.org/TR/rdf-sparqlquery/.

[7] Stanford Parser - http://nlp.stanford.edu/software/lex-parser.shtml.

based on templates, but it is not possible to suggest words not used in the dataset. Finally, there are some works, as MashQL [8], where the process of the creating query is based on a drawing connection between entities that is also restrictive in some cases.

There is also the project IBM Watson [6], which takes a user query in natural language and finds an answer to it. The difference from the previously mentioned works is that Watson uses unstructured documents and looks for an answer to the question in them.

3 Research Methodology and Approach

Based on our previous analysis, we identified the method Panto as the most advantageous and successful method. Our approach, as well as this method, is based on finding out dependencies in the given user query. Dependencies identified by the sentence parser are related with a logical representation of the query. In many cases this representation is similar to the representation of facts in the dataset.

But there are still some situations where the representation in the sentence cannot be directly mapped to the structure in the dataset. One of them is the case when we can find two entities in sentence but there is not direct path between these entities in the dataset. In our method this can be achieved by adding our unique mapping between dataset and user query. Our method, is by this approach, able to identify the path between two entities which are not directly connected. Some users of current web use keywords to express their needs in quicker way. By our mapping method we are also able to translate these types of queries.

We can divide our method into two phases:

– Preprocessing phase,
– Query phase.

As we described in the analysis, the methods which use the preprocessing are more accurate in translation of users queries. Thus our method thus uses preprocessing as the initial part of the whole process.

Concerning the vocabulary - we had in mind that it would be advantageous if the user could use his own vocabulary, thus we included this requirement into our approach. The vocabulary problem is definitely a challenge on bigger datasets, what led us to review methods used for this purpose. We present our method for collecting the information and also the weighting system which helps us to understand the user query.

3.1 Preprocessing Phase

Preprocessing needs some time and resources, but on the other hand it brings a benefit in form of the reduced time needed in the translation phase. In our method, the output of the analysis is the lexicon with weights of relevance between words in the dataset and the found alternative words, which we name descriptors. These descriptors are loaded from three main data sources:

– given dataset,
– database of words (e.g. WordNet[8]),
– Wikipedia article names and anchor texts.

Dataset analysis method based on words database. The dataset analysis is focused on all classes, properties and values, from which we use the name and the *rdfs:label* attribute; these are used as descriptors for given entry.

We observed that the dataset itself does not contain the sufficient amount of alternative words and thus we engaged additional database of words (in our case WordNet, which contains also the relations among the words). This brings us the possibility to load more descriptors for every element in the dataset. We also get the descriptor related to the entry based on its parent classes.

But when we have a lot of descriptors, there is a possibility that two dataset entries get the same descriptor, therefore we implemented the weighting system based on the type of descriptor. System is based on the formula for weight calculation of the given word Eq. 1.

$$W_{descriptor} = w_{descriptor\ source} \times \left(w_{upperclass\ const} \right)^{nesting\ degree} \tag{1}$$

The output of this calculation is descriptor weight stored in the lexicon $w_{descriptor}$. Its value is based on the weight of the descriptor source (label, external dataset, object name etc.), the weight of the nesting constant $w_{neting\ const}$ and the current nesting degree (count of the parent class from the main class, for which is the weight calculating). A pseudocode of whole process of collecting descriptors can be found in the following program code.

```
function getDescriptors(obj, cnw)
  descriptors ← {}

  descriptors.insert(obj.name, NAME_WEIGHT_CONST * cnw)
  descriptors.insert(obj.label, LABEL_WEIGHT_CONST * cnw)

  wn_descs ← getThirdPartyDescriptors(
              descriptors,
              WN_CONST * cnw
            )
  descriptors.addAll(wn_descs)

  cnw = cnw * NEST_WEIGHT_CONST
  if (cnw > STOP_WEIGHT_CONST) then
    desc ← GetDescriptors(obj.parent, cnw)
    descriptors.addAll(desc)
  endif

  return (descriptors)
endfunction
```

---

[8] WordNet - https://wordnet.princeton.edu/.

We defined the constant *STOP_WEIGHT_CONST* to stop nesting in the point, where the weight exceeds the given threshold. We also performed several experiments to verify settings of all used constants in the pseudocode and also in Eq. 1, which will be described later. In the example mentioned in the first section of this text we used the word form *birth place*, which is in triples mapped to *dbo:birthPlace* predicate. This mapping can be done due to our lexicon with alternative forms. Also the entity *dbo: Physicist110428004* can be mapped to the word *physicists* thanks to the lexicon, which contains predicate *rdfs:label* with value *Physicist*.

Alternative descriptors. Our method can successfully deal with phrases composed of two or more words. These denominations are often used in properties like *has-Author*, *is-Married* etc. but also in some classes and values like *production company, submission name*, etc. Building on this observation, we divided this problem into two groups:

– prefixes and suffixes seemingly without meaning,
– word phrases composed of two or more words.

The first group concerns the phrases containing words like *"has"* or *"is"*, which carry no meaning. Therefore, it is difficult to find suitable alternatives for them. In these cases, we substitute only the meaningful words. For instance, the property *has-Author* will get alternatives *has-Writer* and *has-Scriber*. Then if the user writes question and uses word writer in it we can identify the right relation more easily. Weights of alternatives are calculated in the same way as described in the program code mentioned above.

The second case, phrases can be resolved in common way, we find alternatives for all words and create new descriptors; then we calculate their weights. The resulting weight depends on the number of replaced words in the phrase and also on the source impact.

Wikipedia article and anchor texts source. The process described in previous text is advantageous in case of properties (like *has-Author*, *birthplace*, etc.) and base entities (like *Person*, *Car*, etc.), but there are many entities and values which describe people, buildings or historical events, where it is not possible to find alternative words in this way. For purpose of getting the descriptor for these types of dataset entries we use Wikipedia[9]. Wikipedia contains a lot of articles, which incorporate many relations between each other. For example, if one article describes parachute, it could contain reference to the parachute inventor Stefan Banic. All references in Wikipedia are included in the text; therefore, it is possible that anchor text refers to another article using different words than the article itself.

One such example can be the text U.S. Army, which is connected to the article about United States Army. It is just synonym or alternative name for the same object; this structure is presented in Fig. 1. By collecting all of these synonyms from many articles we can extend our lexicon by descriptors, which can't be generated by using approach described in previous chapter.

---

[9] Wikipedia - https://www.wikipedia.org/.

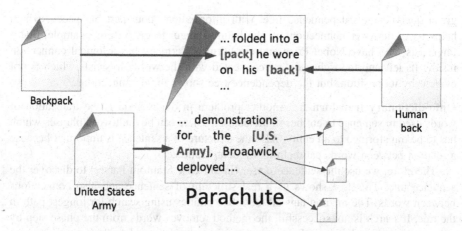

Fig. 1. Structure of Wikipedia articles graph. References are made between articles, using anchor texts with alternative names.

3.2 Translation of Query in Natural Language to SPARQL Query

The schema of our method, which can be found in the Fig. 2, shows steps performed to translate the given natural language query to the structured SPARQL query. The picture is divided into two sections, the rectangle on the right side describes preprocessing mentioned in the previous section and left side of the picture represents the process of translation query itself. The connection between data from preprocessing and our method is Phrase Mapper component, it is responsible for mapping single phrases to its entities in dataset.

The process of translation is highlighted by the darker path with the start in the user interface by submitting a query, then it continues through four main components of our method:

- Query preparation,
- Onto-dictionary transformer,
- Ontology mapper,
- SPARQL transformer.

Our question from the first section: *Which German cities are birth places of physicists with Nobel Price?* is processed by using the Query preparation component and it is transformed to the dependency tree, by using the sentence analyzer. The Onto-dictionary transformer component method maps the words and phrases from the question into ontological elements. Ontology mapper is the most important component of our method, it is responsible for mapping the dependence tree with ontology candidates into a query skeleton. Last component - SPARQL transformer takes the query skeleton and translates it to a SPARQL query, which can be executed on a given dataset.

Query preparation. The natural language question is processed by sentence analyzer; for this purpose, we use Stanford Parser in our method. The sentence analyzer based on given query creates dependence tree, with information about part of speech, which helps us to discover connections in the natural language query. For example, query Physicists who have Nobel Price and are born in Germany has a logical connection inside. Its left and the right side are connected with the word physicists which is not evident by its position, but the dependence tree shows these connections.

Onto-dictionary transformer. Another problem in our method is the translation of words used in sentence to entities in the dataset. There can be multiword phrases which has to be transformed to an entity. Also the structure of a sentence is important because a relation between words can help us in identification.

Therefore, we use the sentence parser (in our case Stanford Parser) to discover the sentence tree. This tree shows us a real structure of sentence and also connections between words. The method tags all identified entities using search for longest path in the tree, If search is not successful, the method removes words from the phase step by step until mapping between the sentence and the ontology is found.

By applying this process, we are able to find many candidates for words in the dataset, and we can identify the best match to given query. All these candidates have their weights which are computed based on weight of the given phrase in the lexicon. This weight is favored if the candidate consists of more words from the query, higher weight has longer phases. For example, from the previously mentioned query it is possible to find word Nobel and also the word Price as a single word in the dataset. However, its joined phrase has higher score and based on it is finally used phrase Nobel Price.

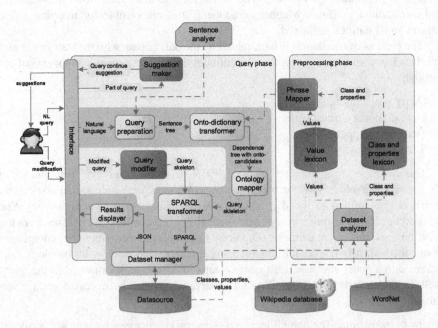

Fig. 2. Diagram describes how our method for translation of natural language query to SPARQL language works.

Ontology mapper. The generated dependency tree has many candidates for every word in it. By weights of these candidates we can find the best scored candidate for translation. One top query candidate is given to the Ontology mapper, where the output is Query Skeleton; The Query Skeleton is structured representation used in our method for the SPARQL query.

To process the query from the first section: *Which German cities are birth places of physicists with Nobel Price?*. The entity *dbr:Nobel_Price* is connected to the *?person* variable by one connection entity. This can't be discovered in the user query, because the user has no knowledge about the dataset structure. But the Ontology Mapper component of our method is able to find this connection. Mapping starts from the best scored candidate, which is used as the central point of the method. Then the mapper continues by connections in dependence tree and looks for the connection with its neighbors.

This is also done in case of the variable *?person*, there we discovered that the variable has *dbo:birthPlace* parameter and also connection with *yago:Physicist110428004* entity. The mapper tries to find the shortest path in the ontology graph between the entity *yago:Physicist110428004* and the variable *?person*, which leads through element *dbo: Nobel_Price_in_{field}* where parameter field can be physics, peace etc. Using this we are able to discover and add these triples to the query skeleton: *?city dbo:country dbr: Germany, ?person dbo:birthPlace ?city, ?person dbo:type yago:Physicist110428004, ? person dbo:award ?award, ?award dbo:subject ?dbr:Nobel_Price.*

4 Results and Discussion

As we described in the previous text our method has two main points: preprocessing and translation, thus we proposed an experiment for each of these points separately. For preprocessing, our own method for weighting used in a lexicon, we firstly prepared series of experiments and based on the results we defined parameters in weighting system.

Then we prepared an experiment to confirm this definition, where 17 people had to order synonyms from high relevant to low relevant for 150 terms from the dataset. In this experiment we had 1008 orderings, and using them we were able to setup and confirm constants used in the method. Using defined parameters our method facilitated to compute right order of synonyms in 89.3% of cases - see Table 1, whereby only orders with the coincidence were counted at every term positions.

The main part of our method is the translation process; we divided it into two steps: translation to onto-dictionary and translation to the SPARQL query. We used NL-SPARQL [3] dataset which contains SPARQL query and corresponding query in natural language; from this database we used 50 queries. These queries were supplied into our described method (V2) and also to our previous method (V1) which not contain mapping to dataset phase.

Translation to onto-dictionary was checked against query form database; we achieved F1 score value 0.78 in this step. Translated SPARQL queries was checked against queries in the dataset. In case of difference between generated and benchmark query we used two assessors which had to judge if the query is translated correctly.

Table 1. Results of ordering experiment

Match	Orderings
Exact	134
1 mistake	10
3 mistakes	6

Table 2. Result of first stage of translation, to triples

	v1	v2
Triples	241	241
Right	100	180
F1 score	0,436	0,78

Table 3. Results of translation

	v1	v2
Questions	50	50
Queries	19	39
Satisfactory	14	36

By our method (V2) we were able to translate 36 queries correctly. Summary of our results can be found in the Tables 1, 2 and 3.

By analysis of our results, we discovered that our method currently does not offer the translation of queries asking for superlatives (Highest mountain, longest river etc.), what will be the part of our future work.

5 Conclusions

In this paper we deal with Linked Data queries and we described our method for translating the natural language query to the SPARQL language. We have presented the approach for the preprocessing phase where we have created our own method for descriptor popularization from dataset itself, WordNet and Wikipedia.

The query translation and also the suggestions are based on weights which were experimentally tuned to the highest possible accuracy. We have also engaged sentence parser to understand user query, and special approach for mapping sentence parts to dataset entities. We performed experiments which let us assume that the suggested research direction brings perspective findings.

References

1. Androutsopoulosa I., Ritchiea G. D., Thanischa P.: Natural language interfaces to databases – an introduction. In: Natural Language Engineering, (1995), pp. 21–81
2. Bernstein, A., Kaufmann, E., Kaiser, C.: Querying the semantic web with ginseng: a guided input natural language search engine. In: 15th Workshop on Information Technologies and Systems, pp. 112–126 (2005)
3. Hakkani-Tur, D., Celikyilmaz, A., Heck, L., Tur, G., Zweig, G.: Probabilistic enrichment of knowledge graph entities for relation detection in conversational understanding. In: Proceedings of Interspeech. ISCA - International Speech Communication Association, ISCA, pp. 14–18 (2014)
4. Cabrio, E., Cojan, J., Aprosio, A.P., Magnini, B., Lavelli, A., Gandon, F.: Qakis: an open domain QA system based on relational patterns. In: 11th International Semantic Web Conference ISWC 2012 (Posters & Demos) (2012)
5. Damljanovic, D., Agatonovic, M., Cunningham, H.: FREyA: an interactive way of querying linked Data using natural language. In: García-Castro, R., Fensel, D., Antoniou, G. (eds.) ESWC 2011. LNCS, vol. 7117, pp. 125–138. Springer, Heidelberg (2012). doi:10.1007/978-3-642-25953-1_11

6. Ferrucci, D., Brown, E., Chu-Carroll, J., Fan, J., Gondek, D., Kalyanpur, A.A., Lally, A., Murdock, J.W., Nyberg, E., Prager, J., et al.: Building Watson: an overview of the DeepQA project. AI Mag. **31**, 59–79 (2010)

7. Furas, G., et al.: The vocabulary problem in human-system communication. Commun. ACM **30**, 964–971 (1987)

8. Jarrar, M., Dikaiakos, M.D.: MashQL: a query-by-diagram topping SPARQL. In: Proceedings of the 2nd International Workshop on Ontologies and Information Systems for the Semantic Web, pp. 89–96. ACM (2008)

9. Kaufmann, E., Bernstein A., Zumstein, R.: Querix: a natural language interface to query ontologies based on clarification dialogs. In: 5th International Semantic Web Conference, pp. 217–233. Springer (2013)

10. Ramachandran, V., Krishnamurthi, I.: NLION: Natural Language Interface for querying ONtologies. In: COMPUTE 2009, pp. 17:1–17:4. ACM (2009)

11. Rico, M., Unger, C., Cimiano, P.: Sorry, I only speak natural language: a pattern-based, data-driven and guided approach to mapping natural language queries to SPARQL. In: Proceedings of the 4th International Workshop on Intelligent Exploration of Semantic Data co-located with the 14th ISWC 2015 (2015)

12. Tennant, H., Ross, K., Thompson, C.: Usable natural language interfaces through menu-based natural language understanding. In: SIGCHI Conference on Human Factors in Computing Systems, pp. 154–160. ACM (1983)

13. Unger, C., Freitas, A., Cimiano, P.: An introduction to question answering over linked data. In: Koubarakis, M., Stamou, G., Stoilos, G., Horrocks, I., Kolaitis, P., Lausen, G., Weikum, G. (eds.) Reasoning Web 2014. LNCS, vol. 8714, pp. 100–140. Springer, Heidelberg (2014). doi:10.1007/978-3-319-10587-1_2

14. Valencia-García, R., et al.: OWLPath: an OWL ontology-guided query editor. IEEE Trans. Syst. Man Cybern. **41**, 121–136 (2011). IEEE

15. Wang, C., Xiong, M., Zhou, Q., Yu, Y.: PANTO: a portable natural language interface to ontologies. In: Franconi, E., Kifer, M., May, W. (eds.) ESWC 2007. LNCS, vol. 4519, pp. 473–487. Springer, Heidelberg (2007). doi:10.1007/978-3-540-72667-8_34

The Use of Semantics
in the CrossCult H2020 Project

Stavroula Bampatzia[1], Omar Gustavo Bravo-Quezada[2], Angeliki Antoniou[1],
Martín López Nores[3], Manolis Wallace[4(✉)],
George Lepouras[1], and Costas Vassilakis[1]

[1] Human-Computer Interaction and Virtual Reality Lab,
Department of Informatics and Telecommunications,
University of the Peloponnese, 22 131 Tripolis, Greece
{s.babatzia,angelant,gl,costas}@uop.gr
[2] GIHP4C Research Group, Universidad Politécnica Salesiana, Cuenca, Ecuador
obravo@ups.edu.ec
[3] Department of Telematics Engineering, AtlantTIC Research Centre,
University of Vigo, Vigo, Spain
mlnores@det.uvigo.es
[4] Knowledge and Uncertainty Research Laboratory,
Department of Informatics and Telecommunications,
University of the Peloponnese, 22 131 Tripolis, Greece
wallace@uop.gr
http://hci-vr.dit.uop.gr/
http://www.ups.edu.ec/en/web/guest/gihp4c
http://atlanttic.uvigo.es/en/
http://gav.uop.gr

Abstract. CrossCult is a newly started project that aims to make reflective history a reality in the European cultural context. In this paper we examine how the project aims to take advantage of advances in semantic technologies in order to achieve its goals. Specifically, we see what the quest for reflection is and, through practical examples from two of the project's flagship pilots, explain how semantics can assist in this direction.

Keywords: Semantics · Reflection · Cultural assets · History

1 Introduction

"CrossCult: Empowering reuse of digital cultural heritage in context-aware cross-cuts of European history" is a newly started project, supported by the European Union under the H2020-REFLECTIVE-6-2015 "Innovation ecosystems of digital cultural assets" funding scheme. The project aims to make reflective history a reality in the European cultural context, by enabling the re-interpretation of European (hi)stories through cross-border interconnections among cultural digital resources, citizen viewpoints and physical venues.

A. Calì et al. (Eds.): IKC 2016, LNCS 10151, pp. 190–195, 2017.
DOI: 10.1007/978-3-319-53640-8_17

CrossCult will use cutting-edge technology to connect existing digital cultural assets and to combine them with interactive experiences that all together are intended to increase retention, stimulate reflection and help European citizens appreciate their past and present in a holistic manner. Interactive experiences and their narratives are designed around four major principles:

- Raise consciousness about the importance of History
- Tackle the study of History from a multi-faceted perspective
- Approach History not only through the written texts from successive eras, but also through all the traces left by those societies (archaeological remains, iconography, epigraphy, numismatics, architecture, art, etc.)
- Reckon that there are no absolute truths in History, but various possible interpretations of the archaeological remains and contrasting viewpoints

In this work we take a closer look at CrossCult's plan to use semantics in order pursue to its goals. We start in Sect. 2 by looking at the notion of reflection, which lies at the core of the project's goals. The triggering, support, evaluation and quantification of reflection are all wide open research directions, in which CrossCult aims to make a sound contribution over the next three years. In order to practically assess the progress made, the project will be implemented on 4 real-world flagship pilots. Herein, in Sects. 3 and 4, we examine how the consideration of semantics will support the reflection in two of the project's flagship pilots. We close in Sect. 5 with our concluding remarks.

2   The Quest for Reflection

Living in Europe we are surrounded by history and culture. When it comes to physical items there are currently more than 19000 museums in Europe and even more archaeological sites [9]. The majority of European larger cities and and even smaller towns and villages either have or are themselves historical landmarks. And as far as digitized items are concerned Europeana already connects more than 30 million objects from over 3 thousand institutions. In fact, the preservation of cultural heritage is at the very core of the foundations of the European Union; the Lisbon treaty, a constitutional basis of the European Union, states that the Union "shall respect its rich cultural and linguistic diversity, and shall ensure that Europe's cultural heritage is safeguarded and enhanced" [8].

As a result, in the period 2007–2013 alone the EU invested approximately 4.5 billion euros in cultural heritage and related research. Most of these funds were directed towards the preservation of heritage, and thus mainly to actions related to the conservation, digitization and related infrastructures. The aim, of course, has been to achieve long term preservation, and thus much emphasis has been given to the sustainable exploitation of cultural heritage assets [1,2]. More recently it has started to emerge that cultural heritage is not just a matter for sustainable, in other words financial, exploitation, but also a powerful social tool that can help strengthen the connections between European people [3].

In this scope, we are now starting to examine not just how cultural assets can be used to generate profit but also how then can help us understand more about ourselves; about our past, present and future; and about the way we are all alike and all connected. In order to achieve this, it is not enough that people visit museums and archaeological sites and pay a ticket. What is truly desired is that such visits make people think and talk about what they saw and how that relates to themselves, their lives and their closer and broader communities.

In other words, the goal is to use the cultural items and locations as mere triggers, through which more important issues can be raised, in both internal (reflective) and external (communicative) processes.

3   Non-typical Transversal Connections

It would take someone about three months of visiting the Louvre on a daily basis from morning to evening to see every item on display - and a few years if all 380.000 items were put on display. The British Museum is even larger than that and possesses an unimaginable collection of approximately 8 million objects. Yet, most of the museums in Europe do not come anywhere near that. In fact, it is quite common for small museums to put on display almost their entire collections and still not have enough to make for a meaningful visit that lasts for a whole hour. This makes it increasingly difficult for smaller museums to attract visitors and therefore to survive.

In CrossCult we aim to use the same objects in order to support multiple narratives; in this way a small collection will be able to sustain the operation of a museum by allowing the collection to be seen many times, each time from a different perspective. Moreover, inline with the project's overall goal, the identified narratives go beyond the typical level of history presentation (e.g. type of a statue, or its construction date), into deeper levels of reflection, over social aspects of life in antiquity, power structures, etc.

The project's third flagship pilot, focusing on exactly these notions, will run at the Archaeological Museum of Tripolis in Greece, where no more than 30 items are on display and not all are necessarily related to each other; the museum is typically visited in a duration of approximately 20 min.

Of course it is not possible to support whichever narrative using the limited set of items available at the museum. Thus, in order to see what is possible, a the semantic map of Fig. 1 has been developed. In this figure yellow ovals correspond to items in the museum whilst the green and purple rectangles correspond to narrower and broader concepts. Each one of these concepts can be the objective of a museum narrative and an experience can be designed in which a visitor is guided through the museum and shown selected items while the concept is discussed. For example, the visitor could be shown items related to social status while asked to consider whether it would have been possible in that era to have a woman running for a major public office as Hilary Clinton is now.

This map has been developed manually, as a proof of concept and because the size of the museum's collection permits it. Within the project we aim to explore

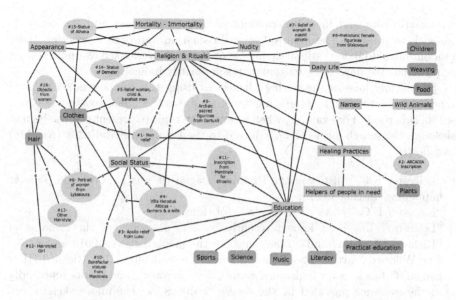

Fig. 1. Semantic map for Archaeological Museum of Tripolis items.

the possibility to generate such maps automatically based on the items' already existing non semantic annotations; in this way a semantic engine will be able to examine a museum's collection and come up with suggestions of interesting narratives that can be designed.

4   Connecting the Past and the Present

It has often been said that museums are not appealing, particularly to the younger generation, because they are boring [4]. In the last decade museums and venues have started to turn to technology to make their presentations more appealing, using tools such as virtual and augmented reality [5]. This approach is great for sustainability, as more visitors are attracted, but it does not foster reflection as the visitors pay more attention to the graphics and the technology than to the content itself. Moreover, as VR and AR become present in increasingly more aspects of everyday life, their presence at a cultural venue is no longer the strong pole of attraction it once was.

In CrossCult we hope to stimulate interest by focusing on the content making it relevant to the visitor's life and reality. Of course this cannot be achieved by static presentations of collections; dynamic, rapidly adaptive presentation methodologies and systems need to be developed, that can be tailored to fit not only each visitor but also each particular day.

To achieve this - i.e. the connection of the venues, the collections and the historical facts to what might stimulate visitors on a particular day - we have made a preliminary implementation of a component that tracks trending topics on social media and identifies random links between those and the concepts we wish to present. The main question in such an approach is whether these random links do indeed stimulate people or they are overlooked as uninteresting or puzzling. We have some extremely positive early indication.

Specifically, one of the authors has been using the component's identified random links to design interesting ways to present content to students in a school classroom. For example, when recently the topic to present was the human skeleton, the search for a (hi)story linked to the human skeleton went (roughly) as follows:

– "Leicester F.C." just won the Premier League (it was a worldwide popular happening on that week).
– "Leicester F.C." plays in "Leicester" (obvious link).
– "Leicester" has been known recently for the discovery of the remains of "Richard III of England" (many news in the media between 2012 and 2015). The Wikipedia article about the "Exhumation and reburial of Richard III of England" has an very high word count for "Leicester", and points repeatedly to the evidence provided by the severe "scoliosis" of the man's skeleton, to techniques of "DNA analysis" for proper identification and to other archaeological/historical evidence.

One can debate whether these links are truly meaningful or not. But the actual classroom experience was clear: the story was planned to be presented in just 10 min, but the pupils' curiosity kept them talking about poor Richard for the whole hour, even though they had never cared about the history of England before.

Thus, through the connection with current events, further discussion has been stimulated. This is a feature that plays very well into CrossCult's goal for the stimulation of reflection. In the scope of the project's second flagship pilot we aim to further explore how a user specific interesting topics - for example topics mined from their recent social media activity - can be used to provide for even more personalized reflection stimuli.

5   Conclusions

In this paper we presented the CrossCult H2020 project and discussed how it will use semantics in order to enhance visitors' reflection on European history and other cross-era, cross-border, cross-culture and cross-gender issues.

We attempted a presentation by example. Through flagship pilot 2 we saw how semantics can provide the discovery of previously unnoticed connections between current events and the bodies of cultural heritage of the presented sites. We also saw how this can spark discussions and stimulate reflection. Through flagship pilot 3 we saw how semantics can help build stimulating presentations of multiple and fundamentally different topics based on only a handful of physical items. Overall, we can see CrossCult as a turning point in the way semantic technologies are used to support not only the exploitation of cultural content but mainly the maximization of its impact through augmented and deeper reflection.

One should of course not oversee the fact that CrossCult is still at its beginning. At the time of writing this text the project kick-off meeting has just taken place and numerous details regarding the project's architecture, methodologies, content, use cases and implementation are still open. Therefore, the specifics mentioned herein should be taken with a grain of salt. Most will turn out as described in the paper, but some will unavoidably be altered as the project's work and research progress.

Acknowledgments. The CrossCult project has received funding from the European Union's Horizon 2020 research and innovation programme. This work has been partially supported by COST Action IC1302: Semantic keyword-based search on structured data sources (KEYSTONE).

References

1. Licciardi, G., Amirtahmasebi, R. (eds.): The Economics of Uniqueness: Investing in Historic City Cores and Cultural Heritage Assets for Sustainable Development, Washington DC, World Bank (2012)
2. Greffe, X.: La valorisation économique du patrimoine. La documentation française, Paris (2004)
3. Dümcke, C., Gnedovsky, M.: The Social and Economic Value of Cultural Heritage: Literature Review (2013)
4. Bartlett, A., Kelly, L.: Youth audiences: Research summary, Australian Museum Audience Research Centre (2000)
5. Lepouras, G., Vassilakis, C.: Virtual museums for all: employing game technology for edutainment. Virtual Reality **8**(2), 96–106 (2004)
6. Bikakis, A.: CROSSCULT: empowering reuse of digital cultural heritage in context-aware crosscuts of European history, ESWC 2016 Project Networking session, 29th May–2nd June 2016, Heraklion, Crete, Greece (2016)
7. Lykourentzou, I., Naudet, Y., Jones, C., Sillaume, G.: Fostering a multi-faceted view of European history: the CROSSCULT project. In: DHBenelux 2016 conference, 9–10 June, Luxembourgh (2016)
8. Treaty of Lisbon amending the Treaty on European Union and the Treaty establishing the European Community (2007)
9. EGMUS European Group on Museum Statistics. http://www.egmus.eu/
10. CrossCult project website. http://www.crosscult.eu/

Author Index

Printed in the United States
By Bookmasters